TECHNICAL
REPORT

T0288943

Freedom and Information

Assessing Publicly Available Data
Regarding U.S. Transportation
Infrastructure Security

Eric Landree, Christopher Paul, Beth Grill,
Aruna Balakrishnan, Bradley Wilson,
Martin C. Libicki

Homeland Security

A RAND INFRASTRUCTURE, SAFETY, AND ENVIRONMENT PROGRAM

The research described in this report was conducted under the auspices of the Homeland Security Program within RAND Infrastructure, Safety, and Environment (ISE)

Library of Congress Cataloging-in-Publication Data

Landree, Eric.
    Freedom and information : assessing publicly available data regarding U.S. transportation infrastructure security / Eric Landree ... [et al.].
        p. cm.
    Includes bibliographical references.
    ISBN-13: 978-0-8330-4031-2 (pbk.)
    1. Terrorism—United States—Prevention—Evaluation. 2. Terrorism—Risk assessment—United States. 3. Transportation—Effect of terrorism on—United States. 4. Transportation—Security measures—United States. 5. Infrastructure (Economics)—United States—Safety measures. 6. National security—United States—Planning. I. Title.

HV6432.L363 2004
363.325'93880973—dc22

                                                                            2006032345

The RAND Corporation is a nonprofit research organization providing objective analysis and effective solutions that address the challenges facing the public and private sectors around the world. RAND's publications do not necessarily reflect the opinions of its research clients and sponsors.

RAND® is a registered trademark.

Published 2006 by the RAND Corporation
1776 Main Street, P.O. Box 2138, Santa Monica, CA 90407-2138
1200 South Hayes Street, Arlington, VA 22202-5050
4570 Fifth Avenue, Suite 600, Pittsburgh, PA 15213-2665
RAND URL: http://www.rand.org/
To order RAND documents or to obtain additional information, contact
Distribution Services: Telephone: (310) 451-7002;
Fax: (310) 451-6915; Email: order@rand.org

Preface

The goal of this investigation was to determine how much data regarding U.S. anti- and counterterrorism systems, countermeasures, and defenses are publicly available and could be found by individuals seeking to harm U.S. domestic interests. The study focused on information that would be freely accessible through Web search and review of library materials. To obtain a reasonably detailed picture of the available information while still covering a range of possible scenarios, researchers examined six different hypothetical terrorist operations involving three categories of transportation infrastructure: air, rail, and maritime. The research team also developed a framework for comparing the amount of information that is publicly available across different terror attack scenarios and infrastructure targets.

The Department of Homeland Security Science and Technology Directorate, Office of Comparative Studies sponsored the study. This report is a response to the U.S. General Services Administration Request for Quotation 41016-Homeland Security Research Studies.

The information presented here should be of interest to homeland security policymakers, and owners, operators, and defenders of elements of the U.S. transportation infrastructure that rely on anti- and counterterrorism defenses for security from terrorist attacks.

This report is one of two under the study "Understanding Terrorist Motives, Targets, and Responses," with Martin Libicki as Principal Investigator. The companion monograph is *Exploring Terrorist Targeting Preferences* (Libicki, Chalk, and Sisson, 2007).

The RAND Homeland Security Program

This research was conducted under the auspices of the Homeland Security Program within RAND Infrastructure, Safety, and Environment (ISE). The mission of RAND Infrastructure, Safety, and Environment is to improve the development, operation, use, and protection of society's essential physical assets and natural resources and to enhance the related social assets of safety and security of individuals in transit and in their workplaces and communities. Homeland Security Program research supports the Department of Homeland Security and other agencies charged with preventing and mitigating the effects of terrorist activity within U.S. borders. Projects address critical infrastructure protection, emergency management, terrorism risk management, border control, first responders and preparedness, domestic threat assessments, domestic intelligence, and workforce and training.

Questions or comments about this report should be sent to the project leader, Eric Landree (Eric_Landree@rand.org). Information about the Homeland Security Program is available online (http://www.rand.org/ise/security/). Inquiries about homeland security research projects should be sent to the following address:

Michael Wermuth, Director
Homeland Security Program, ISE
RAND Corporation
1200 South Hayes Street
Arlington, VA 22202-5050
703-413-1100, x5414
Michael_Wermuth@rand.org

Contents

Figures

Tables

Summary

This report concerns the feasibility of obtaining information relevant to planning terrorist attacks from publicly available sources. To the extent that such information is available, it is particularly valuable to terrorist planners in that it can generally be obtained at lower cost, risk, and effort than more direct forms of gathering information such as observation of a potential target. Familiarity with public sources of information is also valuable to defenders. If they are unaware that a terrorist group knows or can easily learn about a particular vulnerability, that vulnerability can be exploited more easily. If, however, defenders are able to establish a rough idea of what terrorists are likely to know or can learn from public sources, they can better identify what assets, regions, or populations may be at risk and adjust their defenses accordingly.

Given the vast array of information in the public domain, identifying all the information relevant to a potential target and assessing its potential value to terrorist planners is daunting. What is needed is a way to define the kinds of information most likely to be useful in planning and executing attacks on particular targets. We developed a framework to guide assessments of the availability of such information for planning attacks on the U.S. air, rail, and sea transportation infrastructure, and applied the framework in a red-team information-gathering exercise. Our results demonstrate the utility of the framework for identifying publicly available information relevant to planning terrorist attacks. They also allow us to describe the level of difficulty involved in finding various kinds of information relevant to specified attack scenarios.

Research Approach

Our research approach involved four steps. First, we identified six plausible attack scenarios—two each in airline, rail, and sea transportation infrastructures—against which to assess the accessibility of publicly available information. Second, to guide information gathering relevant to these scenarios and to assess the adequacy of results, we developed the modified intelligence preparation of the battlefield (ModIPB) framework. Based primarily on U.S. Army doctrine regarding intelligence preparation of the battlefield (IPB), this framework specifies four categories of information relevant to targets in the transportation infrastructure, including (1) avenues of approach and ease of access, (2) target features, (3) security (including forces, security measures, and other population groups present), and (4) analysis of threats to the terrorist operation. Third, we designated a "red team" to serve as proxies for terrorists seeking information about each of the potential attack scenarios. Team members were instructed to find information sufficient to complete an operational plan for each of the six scenarios, relying on the ModIPB framework as a guide and using only very low- or no-risk information-gathering activities—that is, public source, off-site research. Fourth, we undertook three validation exercises to assess the relevance and completeness of the information collected.

Findings

The primary contribution of this research is the observation that the ModIPB framework is useful in directing analyses of publicly available information that would be needed to plan terrorist attacks across a wide variety of transportation infrastructure targets and attack methods; this outcome suggests that the framework is broadly applicable to the problem of identifying information that might reveal vulnerabilities in those systems. In addition, it became evident from applying this framework what types of information are relatively hard versus relatively easy to find for the set of six scenarios describing potential attacks.

The ModIPB framework is a useful guide to locating information relevant to the planning and execution of terrorist attacks. A detailed presentation of all the results—that is, the kinds of information that the red team did and did not find for each scenario—appears in Appendix A. As a whole, our findings demonstrate that the ModIPB framework performed well as a guide to helping red-team members locate information relevant to the attack. Relying on the checklists we provided, red-team members were able to identify information that, with scattered exceptions, proved useful for planning the hypothetical terrorist attacks across all six scenarios. This assertion is supported by the results of three validation exercises.

Ease of identifying relevant information varied across information categories, with general descriptive information being easiest to find and information concerning detailed security procedures being most difficult to find. Information is considered "easy to find" if, as determined by the red-team exercise, the same type of information is available from multiple sources for multiple infrastructure targets of a similar type (e.g., all airports). Information is considered "hard to find" if only single examples were located or if no information was located. Some types of information could be found for one class of infrastructure or for one scenario, but not others.

Given this variation and the relatively small number of scenarios we studied, we cannot compare the ease of finding information across categories with great precision, but our findings do suggest that certain categories of information are generally easier to find than others. Members of the red team found information concerning the location of terrorist targets, interior structural details, and the size and capacity of security forces relatively easily, but locating information concerning specific security procedures and capabilities was more difficult. A notional summary of the findings is shown in Figure S.1.

For each of the attack scenarios, the red team was unable to locate some of the information that a terrorist planner would need to assess the likely success of a potential attack. For example, for some scenarios, the team found news articles reporting the number of officers that monitor a particular area, but those reports did not provide detailed information about

Figure S.1
Notional Representation of Information Collected by Red Team

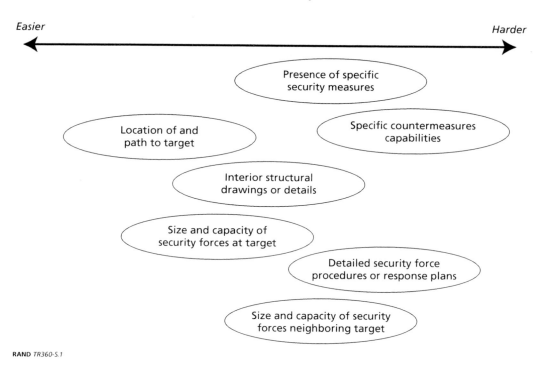

operational plans or deployments at specific stations. That is, the information regarding operational plans and security force deployments was "hard to find."

Policy Recommendations

First, we note that, regardless of how easy or hard it was to locate certain information, there is no evidence from this investigation to suggest that removing information from the public domain would alter the risk of a given scenario occurring. Our findings concern only how easily the red team was able to locate relevant information.

Based on the findings described above, we propose two recommendations intended to help infrastructure owners increase security.

- **To prevent information that includes security details from entering the public domain, review and revise procedures for operational and information security.** Our findings indicate that information pertaining to certain ModIPB categories is not easily accessible through off-site, public information sources. For example, information concerning security force deployments—that is, routes, schedules, number of personnel, vehicles patrolling—is not easily accessible through off-site, public information sources. Nonetheless, our red team did identify a wide variety of kinds of information concerning the air, rail, and sea transportation infrastructures, including overhead images, schematics of sites and equipment, and news reports. Moreover, new information is being added to the public domain every day, along with new capabilities for searching and fusing information. Thus, procedures for securing sensitive information should be evaluated regularly, taking into account developments in technologies for storing and retrieving data, with a view toward identifying vulnerabilities that might allow sensitive information to enter the public domain.

- **Include information that can be obtained from easily accessible, off-site public information sources in vulnerability assessments.** The operations of transportation infrastructure organizations have proven to be attractive targets for terrorist attacks. Thus the owners and operators of these facilities must—and do—conduct vulnerability assessments to identify threats to the security of their assets and activities. To ensure the comprehensiveness of these assessments, information that is appropriately in the public domain must be included.

Our results indicate that the utility and comprehensiveness of information available in the public domain varies by infrastructure and scenario. Given this variation, owners and operators of transportation infrastructure organizations must focus particularly on how information available in the public domain is likely to affect the vulnerability of the specific assets and activities of their own organizations. Relying on ModIPB framework as a tool to guide information searches will help these organizations identify such information, which can then be included in vulnerability assessments.

Owners and operators of transportation infrastructure organizations must determine how frequently vulnerability assessments should be conducted to ensure that, as new information enters the public domain, it is captured in those assessments. Because such new information can enter the public domain at any time, including the day after a vulnerability assessment is conducted, we cannot specify a priori how frequently such reviews should be conducted. We believe, however, that analyses of information in the public domain should either be integrated into current vulnerability assessments or, if conducted separately, should be carried out with at least the same frequency.

Acknowledgments

We would like to thank the infrastructure owners, operators, and subject matter experts who made themselves available for us to interview during this investigation. This research would have been much more difficult without their willingness to share and their frank and open comments.

We are indebted to our sponsor, Robert Ross from the Department of Homeland Security, Science and Technology Directorate, Office of Comparative Studies.

We would like to thank our RAND Corporation colleagues Dave Frelinger, Brian Jackson, Lowell Schwartz, Bruce Grigg, and Michael Wermuth, whose feedback and insight helped contribute to the research direction, findings, and the final document. We would like to thank our RAND colleagues who served as subject matter experts: Russell Glenn, David Mussington, Don Stevens, and Captain Samuel Neill, USCG. We would also like to thank our reviewers for their thorough and insightful suggestions and recommendations.

We thank Maria Falvo for her assistance in helping us complete the written report. Special thanks to RAND communication analysts Susan Bohandy and Jolene Galegher for their writing and organizational efforts, which were invaluable in communicating our findings in this final document.

Abbreviations

ANSI	American National Standards Institute
AS&E	American Science and Engineering, Inc.
COA	course of action
CBP	U.S. Customs and Border Protection
CSI	Container Security Initiative
C-TPAT	Customs-Trade Partnership Against Terrorism
DHS	U.S. Department of Homeland Security
DoD	U.S. Department of Defense
DOE	U.S. Department of Energy
DoT	U.S. Department of Transportation
EDS	explosive detection system
FAA	Federal Aviation Administration
FAS	Freight Assessment System
GAO	Government Accountability Office
GATX	General American Transportation
HEU	highly enriched uranium
IPB	intelligence preparation of the battlefield
IT	information technology
LA/LB	Port of Los Angeles/Long Beach
LAX	Los Angeles International Airport
LIRR	Long Island Rail Road
LPG	liquid petroleum gas
MANPADS	man-portable air defense system
ModIPB	modified IPB
MTA	New York City Metropolitan Transportation Authority

NFPA   National Fire Protection Association
NIST   National Institute of Standards and Technology
NTSB   National Transportation Safety Board
NYC   New York City
OCOKA   observation and fields of fire, concealment and cover, obstacles, key terrain, and avenues of approach
PRD   personal radiation detector
PFNA   pulsed fast neutron analysis
RB-HS   Homeland Security Response Boat
RFP   request for proposals
RIIDs   radiation isotope identifier devices
ROE   rules of engagement
RPM   radiation portal monitor
SME   subject matter expert
TSA   Transportation Security Administration
USCG   U.S. Coast Guard
VACIS   Vehicle and Cargo Inspection System

CHAPTER ONE

Introduction

This report concerns the feasibility of obtaining information relevant to planning terrorist attacks from publicly available sources. To the extent that such information is available, terrorists may be able to obtain it with little risk, as they need never set foot on the site of a potential attack target. With the growth of the Internet, the amount of freely available information—of all sorts—has risen enormously. Google®, for instance, references in excess of 8 billion pages.[1]

This growth has raised questions, particularly since September 11, 2001, about whether sensitive information is too easy to acquire.[2] In addition to increasing the volume of information available, technology has increased the durability of information in that low-cost digital storage and the emergence of digital archive sites have made it more difficult to remove information once it has entered the public domain.[3] The U.S. government and owners of facilities likely to be of strategic value to terrorists (e.g., nuclear power plants) have considered in some detail whether to allow such information to remain in the public domain. Insofar as terrorists are now attempting to attack softer targets—for example, public transportation and commercial enterprises—owners of such targets may need to decide whether to remove at least some of their most sensitive data from the public domain. This research tests the claim that a great deal of information about U.S. security capabilities and vulnerabilities can be discovered from public sources at no risk to the terrorists seeking the information (Thomas, 2002, 2003).

Understanding what information is publicly available about specific targets can help U.S. security forces determine what information terrorists might have obtained without entering the area they are targeting. Defenders might be able to guess what terrorists can learn from on-site reconnaissance by, for example, walking around the facility themselves, but determining what terrorists can learn from off-site, publicly available sources is far more complicated. This study is intended to address that complexity by establishing more concretely what information

---

[1] Our decision to focus on information that could be gathered from public sources was also dictated by constrained resources and real limits on the risks one could expect RAND researchers to take in gathering data. Google is a trademark of Google, Inc.

[2] For instance, from Paul Magnusson and Spencer Ante (2005), we learn,

> One bit of counsel consultants say applies to just about any business: Don't post sensitive information on the Internet. Says Intellibridge Corp. founder David J. Rothkopf, "We could show a company that one of their fuel trucks was scheduled to deliver to a particular site at a particular time, or show them on the Internet blueprints of their most sensitive areas." Utilities, transportation companies, and hazardous materials manufacturers quickly hid such information after audits.

[3] Two contemporary examples of such sites are *Internet Archive* (undated) and Young (undated).

can be obtained from such off-site sources and providing a broadly applicable method for so doing.[4]

Knowing what terrorists know or can learn may be advantageous to defenders. If defenders are unaware that a terrorist group knows about a particular vulnerability, that vulnerability can be exploited more easily. If, however, defenders are able to establish a rough idea of what terrorists are likely to know or learn from public sources and how that information is likely to shape an attack, they can better identify what may be at risk and adjust their defenses accordingly. For example, if defenders are aware that terrorists know the times and location of specific patrol routes, they can adjust their operational plans accordingly to prevent attackers from collecting this information and using it effectively. If defenders know which of their countermeasures terrorists might know about, they can take steps to ensure that these countermeasures remain effective despite having been discovered, or they can shift to alternative defensive strategies.

On the other hand, if terrorists discover countermeasures[5] instituted by security forces, they can take those measures into account in developing operational plans. In particular, the more information that terrorists are able to discover through off-site reconnaissance, the more efficient any on-site reconnaissance is likely to be. If, however, defenders know what information is available only to those who work at or are closely affiliated with the site, what can be gained through legal on-site information-gathering activities, and what can be gained through off-site reconnaissance, they can adjust their security plans accordingly.

This report seeks to improve our understanding about what information may be publicly available about potential targets in two ways: first, by providing an analytic framework for the evaluation of simulated terrorist intelligence collection efforts that can be used for consistent and comparable analyses across scenarios and targets and second, by presenting the findings from a simulated intelligence-gathering exercise (red team) for six specific attack scenarios.

---

[4] Since September 11, 2001, there have been numerous research studies and reports by both the public and private sectors concerning surface transportation security and vulnerability assessments. Examples include reports by the Government Accountability Office, the Congressional Research Service, and the National Cooperative Highway Research Program and Transit Cooperative Research Program of the National Academies' Transportation Research Board. These programs have produced workshops, discussion groups, reports, guidelines, training materials, and vulnerability assessment tools for securing surface transportation infrastructure. More information may be found at National Council for Science and the Environment (undated), Transportation Research Board of the National Academies (undated[a], undated[b]), and U.S. Government Accountability Office (2006). This study addresses a specific issue that has not been emphasized in these research activities: What types of information useful for terrorist planners could be collected through off-site or remote information-gathering activities?

[5] Not everything that is found, particularly if it is a countermeasure established by defenders, is necessarily understood to be a countermeasure; it may simply be seen as an aspect of the target that has no obvious relevance to the operation. The terrorist researcher may discover it and not communicate as much (unless asked to report on it specifically), having deemed it unimportant. This is less likely to be an issue if the owner labels the countermeasure as such, for instance, in a security plan.

Levels of Risk in Information Gathering

For terrorists, the primary deterrent to information gathering, whether on-site or off-site, is the risk of detection. They must always consider the importance of the information to be gained through some information-gathering enterprise in relation to the possibility that finding that information will involve being observed, arrested, or possibly even killed. Moreover, the risk of seeking a particular piece of information is not an objective value, independent of the characteristics of the information-seeker. In particular, terrorists may face more risk in gathering information about a given target than would another individual or group precisely because they may be either known to the authorities or exhibit personal or behavioral characteristics that draw the attention of authorities. Although such factors introduce some imprecision in estimating the risk of a specific information-gathering activity, it is nonetheless possible to categorize forms of information gathering broadly in terms of the level of risk associated with them. Below, we describe the constellations of factors that identify information-gathering activities as constituting no-, low-, medium-, and high-risk information gathering.

Negligible-Risk Information Gathering

Negligible-risk information gathering[6] includes surfing the Web, listening to or watching mass media, reading for-sale material (e.g., newspapers), and perusing information in public libraries. The information that these sources contain has already been recorded, however formally or informally (e.g., Web-logs, or blogs). Much of this material—for example, weather reports, transportation schedules, and maps—is publicly available information.[7] Similar negligible-risk material includes facts that can be learned through casual observation; facts of this sort are what someone can observe without arousing suspicion such as observations from a road, a city street, a park, or as a member of a guided tour. If such information can be collected easily, little security would be gained by removing any such material from the public domain.

Low-Risk Information Gathering

Low-risk information-gathering activities have some potential to arouse suspicion or may entail leaving identifying information behind. Taking a guided tour once may draw no attention, but doing so several times in a relatively short period may arouse suspicion. Security forces may not notice a particular person passing by a point but may take note of those who loiter or who repeatedly return without apparent purpose. Activities carried out during surveillance may also attract attention; photography, for instance, is sometimes forbidden in or around government

---

[6] Strictly speaking, there is no human activity that involves zero risk, and there are ways for defenders to introduce risk even in Web-surfing (e.g., as part of an active defense strategy). Terrorist "surfers" have to watch out for sites that may introduce spyware into their machines capable of capturing information about the user and, thereby, learning something about the user machine's IP address, the keystroke signature of the user, and any miscellaneous telltale information on the user's hard drive. Web sites may also display enticing (even if bogus) information that may tempt those who believe it into revealing themselves. A more comprehensive depiction of countermeasures and counter-countermeasures, and how they affect the risk of gathering data through mass media channels, although possible, is beyond the scope of this report.

[7] For an in-depth examination of the availability of maps and related information see John Baker et al. (2004).

buildings or other properties. Any effort to take photos of such targets is therefore particularly likely to draw the attention of security personnel.[8]

Other forms of low-risk information gathering include monitoring police radios, accessing paid Web services, visiting private libraries, and obtaining information by writing for it or by asking someone in an official position. To monitor police radios, the observer must be within a certain radius of the radio system and being caught with the equipment may raise suspicions; to access Web services, one risks being identified in paying for the service; and visits to private libraries (e.g., those maintained by trade groups) make one vulnerable because, in many cases, identification is needed to enter.

Medium-Risk Information Gathering

This category includes higher levels of information gathering, such as physical surveillance, that require terrorists to establish a presence in, or make repeated visits to, the infrastructure of interest in order to observe it. The hijackers of September 11, 2001, for instance, took several airline trips to various U.S. destinations to satisfy themselves that they could get past security protocols. Likewise, those who bombed the USS *Cole* rented apartments located at the port of Aden to understand the typical vessel traffic at the port. Some techniques, associated more with hackers than with terrorists, include social engineering (i.e., the extrication of information over the phone or the Internet under false premises).

High-Risk Information Gathering

High-risk information-gathering techniques are activities that are likely to draw authorities' attention. Examples include trespassing, planting a monitoring device in a facility (or a long-range listening device near it), computer hacking into highly sensitive or secure sites, acquiring a sensitive (as opposed to, say, clerical or operational) position within a facility for the purposes of inside information gathering, recruiting an insider, or infiltrating a work site. Such activities are more likely to be within the ambit of a state intelligence agency (in part because they require a higher level of resources) than of a terrorist group, but it is possible for them to fall within the ambit of a terrorist organization willing to take risks or one that has access to sufficient resources.

Determinants of Information Gathering

Carrying out a successful terrorist act requires operatives, weapons, money, and information. This last requirement, information, is unique in the sense that so much of it is free or nearly free for the taking, available through the media, in print, or from the Internet. But even though information is freely available from public sources, there is no guarantee that a terrorist researcher will find it.

---

[8]   For several months in 2005, for instance, passengers were enjoined from taking pictures of the New York subway system. Several years earlier, an individual drew suspicion upon himself for taking pictures of a power plant and was arrested and deported. See Democracy Now! (2004).

Information gathering can be complex, with many variables affecting the kind of research that a given group will do and the success with which it does it. Terrorist researchers may confront a vast amount of information housed in a variety of sources, from the Internet to human informants. They must judge what parts of this information are relevant, reliable, and current, given the goals and characteristics of the operation being planned. They must evaluate how accessible each information source is, considering the different levels of risk, different costs, and different levels of effort involved in mining different sources. For instance, for terrorists with high-speed Internet connections, downloading maps may be free of either risk or cost and nearly effortless. In contrast, infiltrating a security organization to investigate its tradecraft is highly risky, quite costly, and involves a great deal of effort.

Here, we discuss three factors that are likely to affect information gathering: target choice, attack-planning stage, and availability of information on the World Wide Web. We note, however, that the relationship between these variables and information gathering is complex. For example, target choice will certainly affect information gathering, but information gathering may also affect target choice. Below, we focus on factors that affect information gathering but acknowledge the possibility that influence may run in the other direction.

Choice of Target

The amount or type of information required to support a decision to attack a particular target depends on the terrorist's certainty about what the target will be. In some cases, terrorists may want information to decide among various targets; others may start with one target in mind; still others may choose targets almost arbitrarily, focusing on whatever opportunities present themselves. Very careful terrorist groups (such as al Qaeda, with its extended surveillance cycle) may require details about security measures at a specific target before they will consider finalizing their operational plan.

Terrorist organizations who choose to carry out a large number of parallel, relatively small-scale, independent attacks (i.e., multiple suicide bombings done by individual terrorist cells) may realize that some percentage may fail or result in members of the terrorist cell being caught. As a result, they may be more likely to assume a higher level of risk in information gathering than terrorists who are putting all of their resources and information-gathering efforts into a single large attack. In addition, a group's willingness to accept a higher level of risk to gather any one piece of information about a target tends to be low if there is a great deal of additional information that would also be needed in relation to the same attack (i.e., if one must make multiple visits to the same target to gather such information). In general, if the point is to scan a large number of locations, low-cost, low-risk approaches may be more attractive.

Stage of Attack Planning

The full range of information needs is almost never obvious at the outset of planning. Information discovered in the early stages of planning often leads to new information requirements. For instance, if investigation of a subway system's security plan reveals the use of bomb-sniffing dogs, many further considerations then arise: how often the dogs are used; where the dogs are used; how they are trained; how sensitive they are; and how they would they react to the presence, for example, of poison gas dispensers. If the discovery of these dogs makes terrorists revise their attack plans, new information may be required to support the new plan.

In particular, the target-selection process may require different kinds of information than does planning the attack. As a result, the terrorist researcher may, over time, carry out multiple information-gathering activities to obtain all the information needed to cull the list of potential targets or waypoints. Once that is done, a more focused search may be feasible. Consider, for instance, a scenario, to be discussed later in this report that entails shipping a container with nuclear material from a foreign port through the Port of Los Angeles/Long Beach (LA/LB). The terrorist would want to find an overseas port where outgoing containers are not scrutinized rigorously, but visiting all ports in the world is infeasible. After culling the most improbable sites, however, visiting the remaining few may be possible. What may start as a no-cost search through public sources may eventually evolve into requirements for more detailed information and even require higher-risk information-gathering strategies. But as new information requirements present themselves, it may be possible that the most appropriate activity is continuing a low-risk and low-cost information-gathering approach, such as a Web search.

Availability of Information on the World Wide Web
Because the World Wide Web offers the possibility of gathering substantial amounts of information at low risk and low cost, it is likely to be among the first sources that terrorist planners consult. Thus, it is important to understand what information terrorists can find there and what is unlikely to be available.

Information on the Web ranges from what is obvious and requires no particular expertise to find to what may be more difficult to find without privileged entrée or special skills in information retrieval. Anyone searching for information and lacking specialized knowledge about the topic is likely to go first to a search engine. Home pages of relevant institutions are another place to visit. Search engines such as Google® are quite efficient, but they do not cover the entire Web. Nonetheless, even lacking training in information retrieval, terrorists may be able to guess which search-word will elicit what material and put it at or near the top of the search result stack.

There are, however, technologically imposed limits to the kinds of information a terrorist might find. For instance, many Web pages contain information that is specifically denied to Web-crawlers (which are used to populate search engines).[9] Other sites are not easily searched by Web-crawlers but have their own search engines; many of these sites contain information that can be accessed only through discussion groups or forums or with special permission. A great deal of information is also generated solely though query results (e.g., airlines schedules). There are also tens of millions of bloggers whose sites are not easily accessed or included in some Web-crawlers.[10]

---

[9]   One notable example was the archive of physics paper preprints assembled at Los Alamos National Laboratory (undated).

[10]   The consolidation and integration of Internet services to a decreasing number of providers (e.g., Microsoft® integration of search tools into its basic operating system, purchase of Blogger® by Google®), along with improvements in search technology, will help to cross some of these boundaries in the future. However, even as tools and strategies for integrating and

In addition, the Web contains sites that are frequented only by those who know some "secret handshake." The best example of communities that use such information-gathering techniques is the computer hacker underground where cliques exchange tricks, techniques, and, sometimes, part of what they have stolen from other people's computers. Similar communities have been described for "superpatriots" (right-wing anti-government activists), as well as individuals affiliated with other terrorist movements.[11] That said, there are sites that do discuss vulnerabilities (e.g., cryptome.org), but they do so in the spirit that motivates "white hat" hackers to reveal flaws in the hopes that identifying them will lead to prompt patching.[12]

However, constraints on the access to certain information can change over time. One should consider the potential emergence of entirely new sorts of information or new ways of accessing that information. For example, the posting and sharing of videos on sites such as youtube.com has surged over the past months (Liedtke, 2006). Although search tools to locate such videos are still primitive, they are likely to increase in number and sophistication over time, providing additional mechanisms to find potentially useful information.

Thus, to the extent that "availability of information on the Web" influences the information-gathering efforts that terrorists undertake—or the results of those efforts, defenders must take into account changes in the kinds of information available and in mechanisms for locating that information.

Information in the Public Domain: How Much? What Kind?

During our investigation, we were never certain whether the inability to find a security countermeasure indicated that no countermeasure was there. We imagine that "real" terrorists face a similar problem, and anything done to exacerbate this problem for them is a security measure in and of itself. However, we recognize that there is pressure on those in charge of securing various components of the U.S. transportation infrastructure to reassure the public that it is well protected. This necessity leads to publicizing security measures and countermeasures. Publicizing such measures may well increase public confidence, but it does so at the potential

---

searching multiple classes of information evolve, new types of information will emerge and be added to the public domain (e.g., more streaming video or camera phone photos). The availability of these technologies will ensure that, even as new capabilities to search across exiting types of information emerge, as new types of information are added, new search strategies and tools will be required.

[11] Our research did not uncover any terrorist sites in which specific physical vulnerabilities were discussed in the same way that hackers trade secrets on the vulnerabilities of, say, Microsoft Windows. We may speculate that such sites do not exist or that, if they exist, they are comparatively rare. There are several possible explanations for the absence of such sites, including (1) hackers are more adept at and comfortable than are random terrorists in setting up Web sites to discuss potential targets' vulnerabilities; (2) the number of potential cyber vulnerabilities is relatively low, permitting focused discussion of them, while the number of physical vulnerabilities is very high, which works against two random individuals having a conversation about them; and (3) the penalties for being caught lurking on hacker Web sites are much lower than similar penalties for being caught lurking on terrorist Web sites.

[12] There is active debate within the computer hacking community about the ethics of revealing flaws in computer software without giving the vendors time to introduce a patch. To get a flavor of the debate, see Ross Anderson's (Professor of Security Engineering at Cambridge University's Computer Laboratory) undated Web page on economics and security.

cost of providing terrorists with information that they need to know to plan their attacks. Worse still, once a countermeasure of a certain type is known to be in place, terrorist information-seekers can search for additional details on the standards for and construction of such countermeasures and, if they are successful in locating that information, attempt to find or generate counter-countermeasures.

In some cases, publicizing security measures may dissuade terrorists. Consider the following hypothetical scenario: Assume that every U.S. passenger plane is equipped with a missile defense system that would protect the plane from shoulder-fired rockets. If terrorists know that every U.S. passenger plane is equipped with some kind of missile defense system, they may consider the target too difficult to attack and, instead, move to less well-protected targets. Such systems may be expensive to implement solely for the purpose of dissuasion. Other countermeasures that may have similar dissuasive effects and that do not require large expenditures could be more broadly publicized. For example, publicizing the fact that an estimated less than one in 10 man-portable air defense system (MANPADS) shots (Stevens et al., 2004) are likely to bring down a plane could produce the same effect at lower cost.

There is a wealth of information available about transportation infrastructure targets, much of it from multiple sources. Information about security forces and countermeasures is considerably slimmer. However, information about security forces and measures can serve several roles. For members of the public, information about security forces and measures can inspire confidence in safety of transportation systems. For would-be attackers, public information about security can have paradoxical effects. On the one hand, it can help them plan operations; on the other hand, it may deter the execution of those operations. Finding a proper balance and deciding what "should" be publicly available remains a challenge and is beyond the scope of this report.

Policymakers and infrastructure owners and operators must also know not only what information to defend, but also what information can be defended. Knowing what information can be defended requires knowing something about what information can be collected easily and what information is more difficult to collect, a central issue in this study. By way of illustration, consider three different bounded sets of information. The first set is defined by what terrorists can learn from off-site reconnaissance, the type of information-gathering activities that are the focus of this investigation. This is the information that is the most difficult to defend, if, indeed, it can be defended at all. The second set of information is what terrorists can learn by on-site information-gathering activities, such as reconnaissance or surveillance, and may involve a higher risk of detection to the terrorist information-gatherer. Defense remains difficult, though there is at least an opportunity for defenders to recognize terrorists in the act of gathering information on-site. Finally, the third set consists of information that is on-site "employee information," that is, information available to those who are employees of or closely affiliated with the infrastructure itself. Employee information would include both public information and insider information. A Venn diagram of these overlapping sets of information is presented in Figure 1.1. Note that the relative sizes and overlap of these different sets of information depends on many factors, such as the specific responsibilities of the employee and the skill, motivation, or goals of the information gatherers.

Figure 1.1
Notional View of Information About a Target

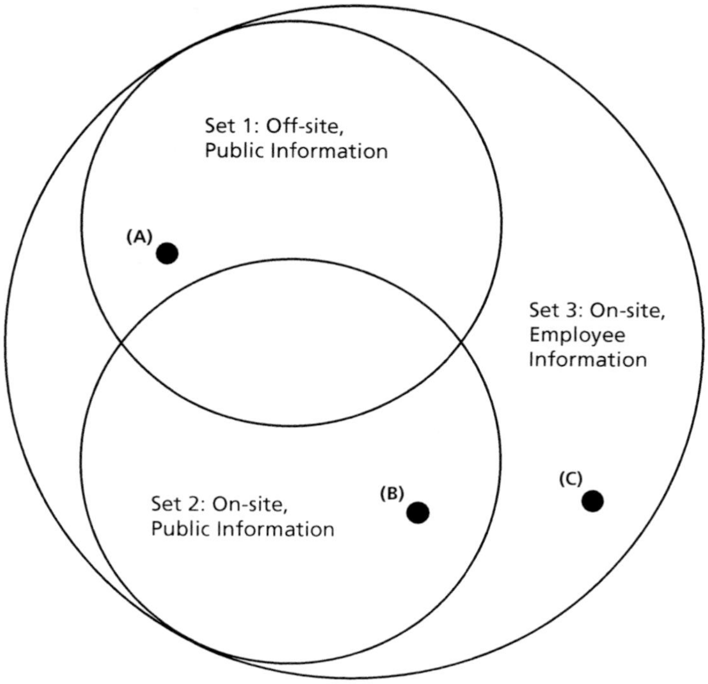

Set 1: Off-site,
Public Information

(A) ●

Set 3: On-site,
Employee
Information

Set 2: On-site,
Public Information

(B) ●

(C) ●

RAND *TR360-1.1*

Whether a given data element falls in one or more of these information sets depends on the characteristics of the infrastructure. For example, as the discussion of our results in Chapter Three will show, information about the existence and use of surveillance cameras was found in off-site, public sources for several of the terrorist scenarios considered and, therefore, falls within the first set of information (point A in Figure 1.1). However, information about the exact location of those cameras, which could be important for planning a terrorist operation, may only be present in the second defined set of information (i.e., on-site, public information), which does not overlap with the first set. Employees will also likely know (or could know if they chose to seek it) the locations of security cameras, so this information would be located in the intersection of the second and third sets of information (point B in Figure 1.1). However, if security cameras are hidden from public view, information about their locations may reside only within the set of information defined as "on-site employee information" (point C in Figure 1.1). Information that can only be found in the third set (on-site employee information) and in neither of the two sets is insider information.

Once policymakers or infrastructure defenders have a good idea of the kinds of data in each of these information sets, they can decide more readily which information to try to keep secure. Employees have access to these data and, in many cases, are obliged to protect them. This report should give those charged with infrastructure security and relevant policymakers a good idea of the shape and general contents of the off-site, publicly available information set for selected scenarios and elements of the U.S. transportation infrastructure. By understanding all three sets and their intersections, defenders can see what information is available only in the insider data set and take steps to protect it—either through efforts to prevent terrorists from gaining access to insider information or through choosing to keep that information from entering the publicly available information set in the future.

Assessing the Results of Information Search: How Much Is Enough?

In the end, the terrorists must assess the adequacy of any information collected themselves. Do they have enough information to proceed confidently? Are they willing to proceed anyway? Have they identified contingencies for countermeasures and other features of the environment that they think they understand well, but are not absolutely sure about? And, perhaps trickiest of all, have they asked all the right questions, or will they be confounded by a forgotten contingency?

The conversion of information into action is a subjective process; confidence resides ultimately within the mind of the terrorist. In some cases, terrorists will know what information they are missing (e.g., they may know that cargo is scrutinized by some criteria, but be in the dark about *what* these criteria are). In other cases, there may be some uncertainty over the extent or importance of missing information. The greater the doubt, the more terrorists are apt to favor simple strategies with multiple redundancies built in—if they go ahead with their plans at all.

Attacks on the Transportation Infrastructure: Six Scenarios

In the previous sections, we have discussed the conceptual basis for our research; in this section, we begin the discussion of our investigative approach.

Recent terrorist attacks—the attacks of September 11, 2001, the attack on jetliners leaving Mombasa, the Madrid train bombings, and the London mass transit attacks—have shown that terrorist groups often favor elements of transportation infrastructures as the targets or instruments of large-scale terrorist attacks. In recent years, a number of U.S. government reports, independent studies, and news stories have openly discussed vulnerabilities and possible scenarios for attacks on transportation targets.

This study is designed to determine what kind of information terrorist researchers with a range of skills, expertise, and guidance can find about specific targets in the U.S. transportation infrastructure from sources in the public domain. As a context for the information-gathering exercise, we use a scenario-based approach, positing six hypothetical operations involving targets within the airline, rail, and shipping sectors of the U.S. transportation infrastructure.[13]

---

[13] Each of these six scenarios has been discussed in the public sphere in news articles and government reports. Five of the six are analogues to attacks that have already occurred.

In each case, we drew our hypothetical scenarios[14] from the public literature but provided a specific locus for our researchers to investigate by associating the hypothetical operation with an actual facility (indicated in parentheses).[15]

Scenarios for Attacks on the Rail Infrastructure

1. Scenario 1: A poison gas attack (NYC subway) (Soares, 2001; Howell, 1998; Japan-101 Information Resource, undated; Council of Foreign Relations, 2004; Online Forum, 1998; Staten, 1995)
2. Scenario 4: Madrid-style bomb attack on commuter train (NYC East River Tunnel) (Biden, 2005; Dateline D.C. Column, 2005; U.S. Library of Congress, 2004)

Scenarios for Attacks on the Air Infrastructure

3. Scenario 2: Bomb in a passenger plane cargo hold (Los Angeles International Airport [LAX]) (Frank, 2003; Air Safety Week, 2004; Epstein, 2003; Donnelly and Novak, 2003)
4. Scenario 5: MANPADS attack on an inbound flight (to LAX) (U.S. General Accounting Office, 2004; Frank, 2003; Ho, 2003)

Scenarios for Attacks on the Sea Infrastructure

5. Scenario 3: Shipping a nuclear device in a cargo container (Port of Los Angeles/Long Beach [LA/LB]) (RFID Journal, 2003; see also Willis and Ortis, 2005; Flynn, 2004; and U.S. House of Representatives, 2004)
6. Scenario 6: Suicide boat rams a docked cruise ship (LA/LB) (Mineta, 2002; Buxbaum, 2004; Roboto, 2001)

With scenarios specifying attack targets and attack modes, we developed a framework for determining what information terrorists would need to carry out these plans. Rather than inventing terrorist information requirements, we used U.S. Army doctrine for intelligence preparation of the battlefield (IPB) and the al Qaeda manual,[16] which is the closest thing we have to terrorist "doctrine," and, in consultation with subject matter experts (SMEs), derived

---

[14] These scenarios overlap with those of the National Planning Scenarios, notably scenario 1 (nuclear detonation, although our scenario stopped once the nuclear device entered the United States), scenario 7 (chemical attack, nerve agent), and scenario 12 (explosives attack, bombing using improvised explosive devices).

[15] This study looks at supply-side factors affecting the selection of terrorist targets. A parallel RAND investigation, "Exploring Terrorist Targeting Preferences" looked at demand-side factors (i.e., what objectives attacking a class of targets might satisfy). The two projects teams interacted and shared one member.

[16] This refers to a translation of a manual dealing largely with security issues, captured in Manchester, England, in the year 2000. It has been hosted on the U.S. DOJ Web site. (See Disastercenter.com, undated.) The quote is from "Eleventh Lesson: Espionage, (1) Information-Gathering Using Open Methods."

a modified IPB (ModIPB) framework from which to choose relevant information requirements (see Chapter Two for details). The ModIPB framework includes four general categories
of information: information related to the approach to the target, characteristics of the target
itself, information about security, and possible threats to the overall success of the operation.

We then designated a red team of researchers to serve as proxies for terrorism researchers
to investigate how much of that information—including data on countermeasures instituted
by security forces to protect these targets—could be found from public sources on the Internet
and in public libraries. The composition of the project red team is consistent with instructions
in the al Qaeda manual, which indicate, "The one gathering public information should be a
regular person (trained college graduate) who examines primary sources of information published by the enemy (newspapers, magazines, radio, TV, etc.)." With this in mind, we selected
a group of research assistants (RAs) employed by the RAND Corporation, all of whom grad-
uated from universities in the United States and were, at most, casually familiar with the
selected targets. The red team members may have used the mode of transportation identified
in our specific targets on rare occasions or made multiple trips on similar modes of transportation at other locations, but they did not possess any knowledge regarding detailed engineering
or security practices associated with those transportation infrastructures.

We presented team members with the scenarios and the ModIPB framework and asked
them to find information from such sources as the Internet and public libraries. In this way,
we sought to replicate, as well as we could, how a hypothetical terrorist group would search for
relevant information using "regular people." To determine whether there was information that
the red team could not find within publicly available sources, but that is nevertheless available
publicly, we included three validation efforts. First, RAND SMEs familiar with security and
counterterrorism efforts for the transportation infrastructures that appeared in our scenarios
inspected the information found by the red team. Second, we compared what the red team
found with information collected during interviews with owners and operators of transportation infrastructure organizations regarding their security forces and security measures. Our
third validation effort focused on the information-seeker's methods; to test the adequacy of
information search, we asked a researcher considered to be more expert[17] in gathering information than the members of the red team to conduct the same exercise, using a subset of the
questions given to the red team as a guide.

Our methods do not allow us to rule out the possibility of false negatives—that is, the
possibility that information was not found even though it was publicly available—but these
validation efforts decrease the likelihood that the existence of false negatives has distorted our
results. And, by examining differences between the information collected by the members of
the red team and the information collected or presented by the various experts described here,
we can estimate the size of the gap between what exists and what was found.

---

[17] There is more to expert information-searching than simply knowing how to use the Web. A technical expert may be able
to infer more from material—whether it is significant, what else needs to be known about it, what it means—than a novice
can even if the expert is equally skilled at finding it. An expert may also know more information prior to having done any
research.

An Illustrative Red-Team Approach

As with legitimate information-seekers, terrorist researchers likely vary in skill and knowledge, as well as in cultural and social characteristics—including their tolerance for risk—and in available resources. Each will take varying approaches to gathering information. While many will follow a systematic sequence of steps, not all will do so. And even the most methodical planners may overlook what, in retrospect, might have been a mission-critical piece of information.[18]

For groups that take a logical approach to operational planning, certain types of information gathering are better matched to specific decision points in the planning process. For example, when a group is deciding on an operation, it seems likely that information gathering would focus on issues relevant to choosing targets. Assume a terrorist group begins with a menu of options and, further, assume that the group ranks each option on a set of criteria—for instance, the cost of attacking a target or the value to the group of doing so.[19] In such cases, easy access to public information is critical; with so many possible targets, visiting each is most likely both unaffordable and infeasible in terms of time.[20] Furthermore, even though the risk of detection during visits to prospective targets may be low, it is not zero. Unless the group is sufficiently confident that its researcher is not actively being searched for or under surveillance by security officers, they will probably want to minimize site visits. The Internet permits terrorist researchers to investigate a large number of sites at relatively low cost, which is especially valuable in the early stages of planning, before a target has been selected.

As the gains to be realized from low-cost, low-risk, and low-effort research activities decrease, higher-cost and higher-risk avenues requiring more effort may become worthwhile. If, for instance, terrorists find that they need to know how frequently a public venue is patrolled and they cannot find the information through low-risk, low-cost, and low-effort information-gathering activities, they may have to loiter near that vicinity long enough to understand patrol patterns—subjecting themselves to the risk of being detected or captured. New information requirements that emerge later in the planning process may merit renewed low-cost, low-risk, and low-effort research. Overall, the phase of operational planning influences what information terrorists need and, consequently, what they must do to get it and what risks and costs to bear in the process.

There are, of course, critical differences between likely terrorist researchers and the RAND research assistants employed in this study. Terrorists are likely to possess greater motivation, but they may be less familiar with U.S. social, cultural, and economic norms. As mentioned above, the profile of any given researcher will influence the research decisions that he or she makes. Even given the same guidelines, each search may yield different outcomes because of

---

[18] For example, roadwork in the neighborhood of the target might not seem like mission critical information, but becomes so if the planned route to target proves to be unavailable on the day of the attack.

[19] A companion RAND report examines potential hypotheses that may explain why one terrorist group, al Qaeda, may select the targets that it does. See Libicki, Chalk, and Sisson (2007).

[20] In practice, it is unclear how much open-ended research is actually done on a universe of targets in light of the many theories of decisionmaking that hold that people rarely consider a large array of options when making choices. See, for instance, the bounded rationality school associated with Herbert Simon (1976) or Gary Klein's (1998) naturalistic theories.

random differences in search paths. Thus, the goal of our study is to illustrate how the ModIPB framework can be used to determine what information is available to terrorist researchers. We do not claim, however, that relying on this framework to assess the availability of information will allow defenders to reproduce exactly the results that any given terrorist might find. And, again, any information-gathering activity is subject to false negatives—the possibility that relevant information may exist, even though it was not found. Such outcomes may occur because of limitations in the skill of the information-seeker or because the information may be unusually difficult to find, may not be public knowledge, or may not be recorded.

The scenarios permitted the red team to research each of the six hypothetical targets and to present a picture of what a terrorist researcher may or may not be able to find about that target at a low level of risk, cost, and effort. Nevertheless, there are clearly countless other complex systems at risk in the United States, as well as numerous other types of possible attacks on the transportation infrastructure. In designing our study, we set out to create a transferable methodology that can be repeated as part of other red-team exercises for systems or specific targets of potential interest to terrorists. The approach can be applied to assess what information a terrorist group might be able to unearth about other at-risk targets at a similar stage of operational planning.

Overview of the Report

Chapter Two presents a conceptual basis for selecting critical information items by discussing what useful information can be derived from such source material as the U.S. Army's doctrine for IPB and the al Qaeda manual. Chapter Three provides a brief summary of the material that the red team collected for each scenario. Appendix A provides a more detailed description of what the red team found.

In Chapter Four, we describe the availability of certain types of information as identified in the red-team exercise. Drawing on these findings, we portray graphically the sufficiency of this information, in terms of its utility in planning the terrorist attacks outlined in the scenarios included in this investigation. Finally, we present recommendations for policies that reduce vulnerability by preventing certain information from entering the public domain and by evaluating information already in the public domain in terms of its implications for the security of the organization.

CHAPTER TWO
Defining Terrorists' Information Requirements: The ModIPB Framework

What essential elements of information do terrorists need[1] to carry out a successful operation? The answer depends on what decisions need to be made. At the most general level, those questions are, "What should we attack?" and "How should we attack it?" In this report, we have defined "what to attack" in terms of the six scenarios described in Chapter One. Thus the framework for information-gathering that we propose here and the empirical red-team exercise focus on the "how"—that is, on finding information relevant to the practical concerns of executing the attacks specified in the scenarios.

To define these information requirements, we relied on three primary sources. The first was the U.S. Army's methodology for IPB. IPB is a continuous process designed to support military decisionmaking by analyzing the environment and possible threats to military operations within a geographic area. The second source was existing RAND research on adapting IPB for urban operations (Medby and Glenn, 2002). The third was the al Qaeda manual. We viewed all three documents, but particularly the IPB materials, not as instructions for continuously collecting information, but as menus to be scanned for possible information requirements that may prove relevant in different scenarios. Our review resulted in checklists from which to select information requirements for each of the six scenarios.

Drawing from the Army's IPB documentation, we created a list of information requirements specifically for terrorist operations that target elements of the U.S. transportation infrastructure. We then consulted with SMEs in the areas of counterterrorism and military urban operations to ensure that the list characterizes the information a terrorist group might need to prepare an operational plan against such an infrastructure target. In the remainder of this report, we refer to this product as the ModIPB framework. Acknowledging that the range of terrorist operations interests is narrower than those of the U.S. Army, the ModIPB framework, therefore, contains a smaller list of elements than does the Army's IPB methodology.

The ModIPB framework, informed by our review of the al Qaeda manual, allowed us to specify systematically the information needs for each scenario and helped ground these intelligence requirements in legitimate sources. We reviewed information items in the ModIPB for their relevance to a given scenario and then confirmed that these items were consistent with items in the al Qaeda manual. We investigated what information is publicly available for those

[1] This is not the same question as asking what they are likely to look for. The latter is an empirical question and may well include information that is sought to provide assurance and confidence to the terrorists but would not affect any decision made about the operation.

15

items deemed relevant to each scenario. Upon review, for each scenario, a subset of the IPB categories was identified as "critical" or "showstopper" information requirements. A *showstopper* is information that (1) indicates the presence of a countermeasure or capability that could significantly reduce the probability of success or (2) is so critical that terrorists deem that the attack cannot take place if this information cannot be collected. Identification of potential showstoppers for each scenario is included as part of Appendix A.

This methodology is broadly applicable; it can be adapted and replicated by U.S. Department of Homeland Security (DHS) and others to determine what information can be found at varying stages of operational planning about other infrastructure elements that are potential terrorist targets. In particular, the ModIPB framework allows consistent categorical assessments of the sufficiency of available data to plan an effective terrorist operation. It also allows the information gathered by different red teams across different scenarios and targets to be compared and contrasted by creating the opportunity to put them in a common framework. Of course, the results will reflect the expertise and skill of the particular members of the red team. This view also represents the information collected over a discrete period. The quality and quantity of publicly available information changes constantly. Therefore, the results for a particular scenario or target may vary over time.

One limitation of our approach—and, indeed, of any effort to plan a terrorist operation—is the impossibility of specifying in advance which candidate information items will prove to be "showstoppers" in a particular case.[2] For example, the maintenance schedule that has the target closed or minimally populated seems anything other than critical until adversaries find that they have attacked an empty building. Consider how much greater, for example, Pentagon casualties might have been if much of the directly affected portion of the building had not been closed for renovations in the previous months, reopening only a few days before the September 11 attack, or had the hijacked plane hit another, more populated portion of the building.

A second limitation of any information-gathering exercise, including the red-team approach employed in this study, is that a failure to find information about a certain item does not conclusively mean that that information does not exist. As we noted in Chapter One, the researcher may fail to find relevant information because (1) he or she was not expert enough to find the information, (2) the information was unusually difficult to find, (3) it was not public knowledge, or (4) it was not recorded in a recoverable medium. Only after all of these possibilities have been eliminated can one conclude that failure to find information means that it is very unlikely that the information exists.

The al Qaeda Manual

Although the ModIPB is a generic framework for capturing intelligence elements relevant to terrorist operations, it was written neither by nor for terrorists. The closest thing we have to actual terrorist doctrine is the al Qaeda manual, which was obtained by the Manchester, England, police during the search of an al Qaeda member's home. This document offers a wide

---

[2]   As researchers gather information, it could possibly become clear that these things are actually not showstoppers.

range of advice and tradecraft to the would-be terrorist, including discussion of countersurveillance techniques, how to arrange a secure meeting or exchange, how to secure a safe house, and, more relevant to the current inquiry, "information-gathering using open methods."

The manual contains instruction regarding information requirements. It appears in the part of the manual called "Tenth Lesson: Special Tactical Operations" under the heading of "Research (reconnaissance) Stage" and in "Twelfth Lesson: Espionage (2) Information Gathering Using Covert Methods" in the sections labeled "The Description" and "The description of the base or camp."[3] The manual states that "Special Tactical Operations" include "bombing and demolition" of infrastructure targets. The relevant section in this lesson focuses on the characteristics of a target and the surrounding environment. Identifying these characteristics requires answering the 18 questions presented in Tables 2.1 and 2.2.[4]

The "lessons" on "espionage" add the following information requirements, shown in Table 2.3, regarding "bases or camps" to be attacked.

The manual clearly distinguishes between "open method" and "covert" espionage. The "Eleventh Lesson" indicates that at least 80 percent of information about the enemy can be obtained using the public sources identified in Chapter One of this report. The discussion of "covert methods" includes instructions for surveillance on foot and surveillance by car and also

Table 2.1
Exterior Information-Gathering Requirements Described in the al Qaeda Training Manual

Category	Question
Traffic and transportation	How wide are the streets and in which direction do they run leading to the place?
	Transportation means to the place
	Traffic signals and pedestrian areas
	Traffic congestion times
Ingress and egress	The area, physical layout, and setting of the place
Other security risks	Security personnel centers and nearby government agencies
	Nearby embassies and consulates
	The economic characteristics of the area where the place is located
	Amount and location of lighting
	Characteristics of the area around the place

---

[3] Disastercenter.com (undated). The "Research (reconnaissance) stage" section of the "Tenth Lesson" of the al Qaeda Manual begins on the page labeled UK/BM-71 TRANSLATION. The sections labeled "The Description" and "The description of the base or camp" are located on the pages labeled UK/BM-90 TRANSLATION and UK/BM-91 TRANSLATION of the "Twelfth Lesson" respectively.

[4] Interestingly, the "Tenth Lesson" (special tactical operations) discusses information requirements regarding the habits and relationships of individuals as targets for kidnappings or assassinations. The ModIPB framework does not address these requirements, given our project's focus on infrastructure. However, a similar process could be used to develop information categories relevant for other types of terrorist attacks.

Table 2.2
Interior Information-Gathering Requirements Described in the al Qaeda Training Manual

Category	Requirement
Human factors	Number of people who are inside
	Number and location of guard posts
	Number and names of the leaders
	Individuals' times of entrances and exits
Other factors	Number of floors and rooms
	Telephone lines and location of the switchboard
	Inside parking
	Electric box

Table 2.3
Information Requirements Described in the al Qaeda Training Manual About Bases or Camps

Category	Requirement
Human factors	Guard posts
	Unit using the camp
	Number of soldiers and officers
	Commander's name, rank, and arrival and departure times
	Sleeping and waking times (presumably of troops or security forces)
Other factors	Location
	Exterior shape
	Transportation to it
	Space [area]
	Weapons used
	Fortification and tunnels
	Amount and periods of lighting
	Ammunition depot locations
	Vehicles and automobiles
	Leave policy
	Degree and speed of mobilization
	Brigades and names of companies
	Telephone lines and means of communications

explains how to identify potential informants who can provide useful information about the potential target.

The Modified IPB Framework

The basic framework for ModIPB comes directly from the doctrinal U.S. Army IPB. Doctrinal IPB requires intelligence collectors and analysts to

- define the battlefield environment. Identify the boundary of your operational area.
- describe the battlefield effects. Determine how the environment will affect enemy and friendly operations.
- evaluate the threat. Determine the capabilities, doctrine, tactics, techniques, and procedures that threat forces may employ.
- determine the threat courses of action (COAs). Integrate the information from the previous steps to create meaningful COAs.

For our ModIPB framework, we identified four primary categories of information—closely derived from doctrinal IPB—in which to group all possibly relevant information requirements:

- avenues of approach and ease of access
- features of the target
- security (including forces, security measures, and other population groups present)
- analysis of threats to the terrorist operation.

The first two information categories correspond to doctrinal IPB concerns with battlefield environment and effects. The third is drawn from evaluations of the threat, and the fourth is based both on evaluations of the threat and threat COAs.

Each of the four primary categories in the ModIPB contains multiple items. For example, the first category—*avenues of approach and ease of access*—includes elements related to features of the terrain, lines of sight, and accessibility of the relevant parts of the area of operations (see Table 2.4). Doctrinal IPB focuses on maps and various overlays that can be added to display additional information. In any effort to identify clearly the location of the target and available paths to the target, maps distinctly showing the surrounding terrain and buildings are a good place to start. If significant portions of the operation will take place within buildings or underground, blueprints may also be useful. Likewise, if the operation requires breaching doors, windows, or walls or accessing ventilation or electrical systems, it will be useful to know how the building is constructed.

Although a well-planned attack will require forethought about how attackers reach the target, scenarios that require the attacker to be at a specific place at a precise time (such as a coordinated or multiple attack or an attack against a moving target) or that place attackers in a position where they would be identified as such if seen (e.g., overtly carrying arms or

Table 2.4
Elements of the ModIPB Framework (Avenue of Approach)

Element	Details
Target location	
Surrounding terrain and buildings	
	Maps
	Blueprints
	Types of building construction
	"Critical points"
	Observation and fields of fire, concealment and cover, obstacles, key terrain, and avenues of approach (OCOKA)
Available paths to target	
	Exact path(s) to take
	Go/no-go areas (because of barriers, obstructions, or impassable terrain)
	Areas of restricted or limited access (security restrictions)
	Rules or laws governing movement (vehicular and otherwise) in target area
	Traffic conditions (all relevant vehicular and pedestrian modes)

material or undisguised in a restricted access area) require plans by terrorists that include the exact paths they will take. Terrorists planning overt attacks will be more likely to be concerned with factors such as whether they will be observed, presence of physical obstacles, and when or where they are at greatest risk of being seen by security forces (or passersby). Terrorists planning attacks that involve transitions from covert actions to overt attacks (such as a MANPADS concealed in an automobile or boat until ready to be fired) are also likely to be concerned with features that may provide cover or that may be obstacles during the period when the attackers are acting overtly. Attacks that involve point-to-point fires of any sort (again, such as a MANPADS attack) also require information about lines of sight from potential firing positions to the target(s).

The general planning of movement to target, whether such movement is time-sensitive or overt and thus vulnerable to detection, may have to be concerned with go/no-go areas or restricted/limited access areas. Any of the elements under avenues of approach could generate critical points, which are important terrain features such as high-traffic areas, chokepoints, security stations, or locks.

Several items cover on-site movement. Even if the terrain is open and unrestricted, knowledge of the rules governing movement in the target area is important to remaining undetected. Many criminals are apprehended during routine traffic stops, and smart terrorists actively seek to avoid such problems (see the discussion of al Qaeda's intelligence preparation, below). Traffic conditions and variations in accessibility depending on time of day or week can also affect the movement of the potential terrorist in vicinity of the target. If attackers must violate traffic rules or travel in restricted areas (e.g., the last hundred yards of a USS *Cole*–style attack against a cruise ship), they will need to understand how long and how far their travel would deviate from the rules and where the difficulties lie in getting into restricted areas.

Some of the details within the category *target* may overlap with those associated with the *avenues of approach and ease of access* category (see Table 2.5).[5] For example, the potential locations from which to launch an attack may be made clear by a combination of maps and terrain data, along with some of the features and structure (technical details) of the target. Likewise, if the target is mobile, the predictable paths that are used can be plotted on maps. Note that the variability associated with a target encompasses not only movement, but also variations in population character and the density of potential victims in the area. Such variations, along with changes in security and accessibility, can contribute to assessing optimal times to launch attacks against targets.

The third category, *security*, is comprised of three subcategories: *security forces*, *security measures*, and *other population groups*; see Table 2.6. Security forces entail the size and types of security forces (e.g., armed and unarmed, plainclothes, canine, mounted), as well as their jurisdiction. Also of interest are their capabilities (arms and equipment); where they are deployed; the location of their headquarters, stations, and checkpoints; and their behavior as embodied in their security plans, their rules of engagement (ROE) or use-of-force policy,[6] and their response times. Security measures include specific countermeasures, and the use and function of checkpoints, and sensors, including cameras, scanners, and other detection equipment.

Other population groups include the non–security force personnel in the area of operations: The terrorist needs to know who they are, when they are likely to be present, and what they are doing in the operational area. Knowing whether nonsecurity personnel have had training or facility users have been given vigilance instructions that might increase their potential contribution to security may also affect terrorist planning. Knowing regularly scheduled

Table 2.5
Elements of the ModIPB Framework (Target Characteristics)

Element	Details
Attack	Possible locations from which to launch the attack
	Possible times or windows of time to launch the attack
Target	Mobility and variability of the target. If mobile, the (predictable) path it takes
	Relevant features and structure of the target (i.e., technical details)

---

[5] Because our study focused on infrastructures, we have excluded information items associated with targeting specific individuals.

[6] *ROE* refers to the circumstances and limitation under which military forces will initiate or continue combat. *Use-of-force policy* refers to *law enforcement* and *other security forces*; the latter can vary widely from one organization to the next.

Table 2.6
Elements of the ModIPB Framework (Security)

Element	Details
Security forces	
	Locations of headquarters, stations, checkpoints
	Overall size and types (uniformed, plainclothes, canine)
	Number on duty at any one time; hours of duty; variation by times of day
	Applicable operational jurisdictions; number of security forces contributed by each
	Capabilities
	Vehicles and other assets
	Radio frequencies used (and other communications used)
	Rules of engagement or use-of-force policy
	Specific or individual deployments
	Fixed positions
	Patrols (routes, schedules, number of personnel, vehicles patrolling)
	Times of observations (cameras, live operatives)
	Number of security personnel required to be "passed"
	Variations by times of day
	Security plans (operational details)
	Security response plans
	Past performance with previous (similar?) incidents
	Behaviors/plans/capabilities at different levels of alert
	Response times
Security measures	
	Kinds of checkpoints to be passed
	Search procedures (For what will officials be looking or asking?)
	Cameras, scanners or detection equipment operating over the area to be traversed
	Sensitivity of these devices
	Frequency with which sensors are "read"
	Illumination
	Specific countermeasures (e.g., vehicle barriers)

Table 2.6—Continued

Element	Details
Other population groups	
	Other people at the facility (Why are they there? What are they doing?)
	Bystanders, recreational users, passengers
	Differences in population at different times of day
	Vigilance instructions and emergency phones
	Level of security training for nonsecurity employees of the facility
	Schedules of regular arrivals and departures from target area
	Ease of camouflaging yourself as a member of one of these population groups

arrivals to and departures from the target area can help minimize the chance that vigilant citizens observe or interrupt operations. Similarly, knowing which of the population groups offers the easiest or most appropriate route to effective camouflage might be key to maintaining the secrecy of covert operations; for instance, to disperse poison gas or place a bomb in a railroad or subway station, a terrorist may need to be able to pass as a tourist or commuter.

The fourth category is an analysis of obstacles to the success of the operation (see Table 2.7). The elements in this category are assessments based on the information collected in the three categories described above and are intended to reflect an estimate of how that information may affect the terrorist attack operation from the perspective of the information-gatherer (i.e., the terrorist). The analysis asks questions such as, What obstacles do security forces and security measures present? What threat is posed to the operation by non–security facility personnel or by other citizens likely to be present in the area of operations? Again, the overtness and planned mode of operation strongly interacts with the actual capabilities of security forces and other population groups in the area of operations. For example, radiation detectors

Table 2.7
Elements of the ModIPB Framework (Threats to Terrorist Operations)

Element	Details
Threat posed by security forces and measures	
	Deployments, response times, vehicles, equipment, plans
	Cascading information (from organization of oversight and headquarters locations to who will be on the avenue of approach on attack day)
	Estimated effectiveness of security response capabilities (including communications)
Threat posed by employees of the target	
Citizens (concentrations of, heightened vigilance of)	
Weather (as it affects effectiveness of the operation)	

have no impact on scenarios not involving radiological material; baggage x-ray machines have no impact on an operation that detonates explosives before reaching the security checkpoint; well-armed guards at a gangplank have little hope of preventing a suicide boat from ramming a cruise ship.

A final note concerns weather, which is important whenever visibility or navigability matters. Most weather information is publicly available. But more subtle examples that we consider as part of this category, such as the air circulation and whorl and eddy effects in the "microclimate" of an underground subway station, are not easily knowable. When such information is critical to a particular scenario, it can present significant challenges to terrorist planners.

Both the information categories identified in the al Qaeda manual and the ModIPB categories are consistent with those identified in a report from the Transportation Research Board's National Cooperative Highway Research Program, which lists the types of information terrorists or criminals may need prior to an attack (Science Applications International Corporation et al., 2004).

The al Qaeda manual also discusses information requirements that extend beyond the range of scenarios that were considered (e.g., assassinations), and therefore may have a larger scope than that of IPB. However, IPB as a resource provides a more comprehensive and detailed list of information requirements necessary for planning and executing terrorist attacks on physical infrastructures. While the ModIPB and the al Qaeda lists are subject to all the same caveats as any other effort to enumerate intelligence requirements and lack the years of testing and success that doctrinal IPB has as a foundation, they nevertheless constitute an intuitively reasonable list of facts that terrorist operational planners would need to know. A comparison of the categories developed for ModIPB and those that can be inferred from the al Qaeda manual (Tables 2.1 and 2.3) appears in Appendix B.

Moving from Abstract Framework to Real-World Information Requirements

The ModIPB lays out a comprehensive list of the information items that might be needed to plan an attack on a public infrastructure, but not all needs are the same from one group of attackers to another. Neither do they become evident at the same time. The kind of information required at any given time depends on the particular stage of the planning process in which the terrorist is. Any attack plan involves moving through phases that contain key decision points—for example, selecting a target, selecting the mode of attack, choosing specific terrorist operatives, and developing an operational plan. These decisions do not need to be made in a particular order: There is no simple linear outline for planning a terrorist attack. Although it seems reasonable to assume, for instance, that target selection is a decision likely to occur at an early point in the overall process, one could imagine a situation in which a decision about mode of attack might precede target selection, such as when a terrorist organization has come into possession of a MANPADS and wants only to choose a target against which to use it. Similarly, although one can easily imagine a group choosing terrorist operatives for an attack after target, mode, and basic attack plans are already in place, the composition of the terrorist group—that is, who is available to execute the attack and what skills they have—may also influence choices about targets and methods. This is particularly likely if the group is small and intends to plan and execute an attack itself. Table 2.7 details considerations that terrorists might deem threats to their operations.

The red-team exercise that we conducted represents the beginning of detailed operation planning; no information is assumed to have already been collected. Initial information-gathering for operational planning is to be conducted without travel to the target site and through very low- or no-risk information-gathering activities—that is, public source, off-site research. It is reasonable to assume that there may be subsequent information-gathering or planning phases, including higher-risk information-gathering or on-site intelligence efforts, but on-site or similar information-gathering activities are beyond the scope of this investigation. The project red team was instructed to find information sufficient to complete an operational plan for each of the six scenarios, using the ModIPB framework as a guide. Table 2.8 includes a summary of the targets and modes of operation for each scenario.

Table 2.8
Summary of Terrorist Scenario Targets and Mode

Scenario	Infrastructure	Target	Mode of Operation
1	Rail	42nd Street/ Times Square Subway Station	Multiple terrorists positioned around station release gas.
2	Air	Passenger airplane departing from LAX	Explosive device is loaded as cargo into the hull of a passenger airplane.
3	Sea	LA/LB	A nuclear device is transshipped through the port within a cargo container without being detected.
4	Rail	Long Island Rail Road/ East River Tunnel	Explosive devices are planted on a train and set to detonate while the train is in the East River Tunnel.
5	Air	Passenger airplane landing at LAX	MANPADS is launched at a passenger airliner during landing.
6	Sea	Cruise ship at LA/LB	A boat containing explosives is rammed against a cruise ship while docked in port.

CHAPTER THREE

Summary of Red-Team Findings and Validation

This chapter contains a brief discussion of each scenario. We first describe what terrorists might need to know before they carried out the attack portrayed in the scenario and then provide a general description of the results of the information-gathering exercise for each case. In Appendix A, we present detailed examples of the types of information collected by the red team; there, we list the essential elements of information that would likely be needed to carry out a successful terrorist attack in each scenario and catalog the information that project staff were able to find through open sources.

Scenario 1: A Poison Gas Attack on the NYC Subway (42nd Street Station)

Based on analysis of the collected information, critical information requirements for the operation are (1) how to enter and exit the station in the most expeditious way, as well as where to leave the poison gas canisters, (2) air circulation patterns within the station, (3) the density and patrolling habits of security forces, (4) whether gas and other chemical sensors are present, and (5) what other countermeasures are in place. With respect to these requirements, we obtained the following results:

- From the Web sites of construction companies associated with ongoing reconstruction of the subway, the team obtained schematic diagrams of the Times Square Station that included entrances, exits, and the relative distance between the different platforms.
- The team located documents that described the air handling and related systems that could affect the effectiveness of a gas attack in the station.
- The team also located information about the relative size of the New York City Metropolitan Transit Authority (MTA) Police Department and the Transit Bureau of the New York City Police Department Transit, as well as radio frequencies used by the various MTA and Transit Bureau police (RadioReference.com, undated), discussions about the types of suspicious behavior for which police would be looking (Robin, 2005), but no detailed information about specific security plans or procedures.
- News articles were also located that mentioned chemical and biological agent detectors tested for use in the New York City subway and train stations (Smerd, 2005). They provided limited information about the operational use and limitations of these sensors, but no specific details regarding their effectiveness, location, and security response in the case

of an event. There were also reports found by the team that mentioned plans to expand the use of closed circuit cameras throughout the subway system (Luczak, 2005), but no information regarding the specific location or operational use of those cameras.

- The team found no information on any other countermeasures that may or may not exist.

Scenario 2: Bomb in a Passenger Plane Cargo Hold (at LAX)

Based on analysis of the collected information, critical information requirements for the operation are (1) which passenger aircraft carry cargo, (2) the procedures by which cargo is inspected prior to being placed on an airplane, (3) techniques to minimize the adverse inspection of such cargo, and (4) any other countermeasures. With respect to these requirements, the following results were obtained:

- Passenger airline Web sites indicated whether they provided cargo handling services and where such services originated.
- Several media articles discussed vulnerabilities (e.g., highlighting the fact that a percentage of cargo is loaded onto passenger carrying planes without being scanned for possible explosive devices [for example, Schneider, 2002]); others discussed the limitations of current technology and referred to pilot programs to test the screening procedures. Several of the vendors for explosive detection equipment provided product specifications that were available for download, but very few provided details regarding resolution limits or penetration depth. No specific information about security screening processes, procedures, or the technologies used by specific airlines was found.
- A number of Web sites provided instructions for becoming a known shipper with the particular airline, and, in some cases, the conditions under which they would ship packages from unknown shippers (the Transportation Security Administration [TSA] "known shipper" program is designed to prohibit carriers from accepting cargo that does not originate from shippers who are registered and meet the TSA requirements [Public Law 108-90]); what little information we found about the screening process for becoming a known shipper was found came mostly from first-hand accounts of people in newsgroups and online discussion groups (Harford Reptile Breeding Center, undated).
- The team found little information on the use, specifications, or effectiveness of other bomb mitigation strategies such as blast-proof cargo containers and cargo liners.

Scenario 3: Shipping a Nuclear Device in a Cargo Container Through LA/LB

Based on analysis of the collected information, critical information requirements for the operation are (1) which outgoing ports are most thorough in inspecting outgoing cargo, (2) what procedures and equipment are used to inspect incoming cargo, (3) techniques to minimize the adverse inspection of such cargo, and (4) any other countermeasures. With respect to these requirements, the following results were obtained.

- Government Accountability Office (GAO) reports on the U.S. Customs and Border Protection (CBP) Container Security Initiative (CSI) described a program aimed at intercepting high-risk cargo by conducting inspections at foreign ports, before the cargo is shipped to the United States. However, according to one report, only about 17.5 percent of containers deemed high-risk were inspected at the point of departure (Stana, 2005b).
- Media articles described the planned expansion of the use of radiation portal monitors (RPMs) and other radiation detection tools at LA/LB to the effect that it was to have RPM coverage for all incoming cargo by the end of 2005 (CalTrade Report, 2005) but contained little information on the effectiveness of current technology and screening policies. Several documents discussed the methods and standards used for testing and evaluating the performance of radiation portal monitors; the information they provided was helpful in understanding some of the technical requirements. Nothing was found on the performance of specific RPMs. Two government documents, a DHS bulletin (Mayer, 2005) and a 2004 U.S. Department of Defense (DoD) Defense Science Board report, mention performance issues with current radiation portal monitors.
- The search revealed no details about the process that CBP uses to identify high-risk cargo. Several sources mentioned the reliance on shipping manifests in determining high-risk cargo; however, no specifics about the weights, algorithms, or other data elements could be uncovered. Several reports and media documents discussed the effectiveness of the DHS Customs-Trade Partnership Against Terrorism (C-TPAT) program and observed that, for a company or shipper to have C-TPAT certification, it is required to have completed a self-assessment that meets DHS security guidelines. However, according to a 2005 *USA Today* article, DHS has audited only approximately 11 percent of C-TPAT certified shippers.[1]
- The team found no information on any other countermeasures that may or may not exist.

Scenario 4:
Madrid-Style Bomb Attack on Commuter Train in the NYC East River Tunnel

Based on analysis of the collected information, critical information requirements for the operation are (1) schedules sufficiently detailed to suggest when the commuter train would be inside the East River Tunnel, (2) the density and patrolling habits of security forces, (3) the presence

---

[1] Hall (2005). In an interview that was published after the information collection phase of this study, the director of the C-TPAT program states that, as of January 2006, DHS had audited approximately 25 percent of C-TPAT members. See Tirschwell (2006).

of gas and other sensors, and (4) any other countermeasures. With respect to these requirements, the following results were obtained.

- Schedules were easy to find and readily available. Information on how precisely the train operators adhered to these schedules, however, was rarely available.
- Several media articles mentioned the number of police officers who monitor the Long Island Rail Road (LIRR) stations, bridges, and tunnels (Donohue and Gittrich, 2004). One quoted a LIRR official stating that MTA police were monitoring certain high-value targets and mentioned several examples of such targets. Web sites also provided information about police frequencies that MTA police use.
- Several Web sites showed images of the track and stations for the LIRR, but no obvious scanners, cameras, or police presence at the stations was detected in these pictures. Media articles reported that the LIRR was increasing its use of electronic access control along with upgrades to their surveillance cameras and intrusion detection systems.
- According to some of the articles, MTA officials had expressed interest in proposals to prevent cell phones from being used as triggering devices while still enabling cell phone reception (Smerd, 2005).
- The team found no information on any other countermeasures that may or may not exist.

Scenario 5: MANPADS Attack on a Flight Bound into LAX

Based on analysis of the collected information, critical information requirements for the operation are (1) the location of good firing positions, (2) the possible use by airlines of countermeasures against MANPADS, (3) the density and patrolling habits of security forces, and (4) any other countermeasures. With respect to these requirements, the following results were obtained.

- Useful information on the physical area around LAX as well as information on flight paths for incoming and outbound flights was available, but there were very few ground-level images detailing all possible obstructions (e.g., tall buildings) away from the airport.
- Several media articles and reports mentioned the lack of antimissile equipment on commercial aircraft. There was some mention of tests that were to begin in 2005; however, no other details were available. Red-team members did discover a 2004 news report that mentioned increasing the use of helicopter surveillance, fencing, and stepped-up police patrols to protect LAX from missile threats ("Security Boosted at LAX to Guard Against Missiles," 2004).
- Some information about the size of the LAX police department as well as about additional officers from special units was available, but little information on security procedures employed beyond the LAX perimeter was found.
- The team found no information on any other countermeasures that may or may not exist.

Scenario 6:
Suicide Boat Rams a Docked Cruise Ship at the Port of Los Angeles

Based on analysis of the collected information, critical information requirements for the operation are (1) the location and scheduling of cruise ships, (2) blueprints of cruise ships that may indicate particular vulnerabilities, (3) the density and patrolling habits of security forces, and (4) any other countermeasures. With respect to these requirements, the following results were obtained.

- Multiple overhead images and maps showed the location of cruise ships at the Port of Los Angeles, as well as data on their arrival and departure times.
- The team found no ship schematics or technical drawings that revealed details such as hull thickness, the location of bulkheads, or other engineering information.
- Some documents referred to the development and deployment of waterborne perimeter security barriers (Port of Los Angeles, 2004), but they did not provide details on their effectiveness or operational use. Other sources discussed the various security procedures to protect cruise ships while in port, including "no float zones" around cruise ships and U.S. Coast Guard (USCG) escorts (U.S. Code, 2003).
- The team found information about the use of helicopters with sharpshooters normally used against drug smugglers, but this information did not indicate whether such capabilities would (or could) be used to intercept potential suicide boats.
- The team found no information on any other countermeasures that may or may not exist.

Validation

The red-team exercise reported here was subject to three levels of validation: a review by SMEs, a review based on interviews conducted with transportation infrastructure insiders, and an assessment comparing the results obtained by a RAND expert tasked with the same exercise (a red-team expert) to those obtained by the team. This last exercise was used to determine whether the expert would find more or different information than did the red team. Because of resource constraints, it was conducted for only a subset of the ModIPB categories for only three of the six scenarios.

RAND SMEs who had experience with each of the targeted infrastructures were asked to review the red-team findings for comprehensiveness and accuracy. All of the SMEs acknowledged the breadth of material uncovered; none found major errors or problems with the scenarios or the information collected.

Project staff interviewed operators, security personnel, or members of infrastructure associations affiliated with each of the target types. A total of nine interviews were conducted. Because conversations with infrastructure insiders were unclassified, RAND could not be sure whether additional countermeasures that they could not discuss are used at their facilities, but the interviewees were generally forthcoming. The interviews provided the project staff with information about the types of security information that would be considered sensitive or possibly harmful if they were available in the public domain. They also provided the project staff with valuable information for reviewing the findings of the red team. Accordingly, the information collected by the red team was accurate and consistent with what was learned through the interviews and through consultations with other subject matter experts.

For each of the first three scenarios, the material uncovered by the red-team expert is detailed in Appendix A. The expert did find more and slightly different information than did the project red team, but the differences were small and did not suggest the red-team results were inaccurate or incomplete.

CHAPTER FOUR

Conclusions and Recommendations

As indicated in the previous chapter, the red-team information-gathering exercise revealed that a substantial amount of information regarding each of the attack scenarios that we defined as available in public sources. This chapter discusses this finding, explores the extent to which broader conclusions can be based upon it, and makes recommendations concerning the appropriateness of DHS actions or decisions.

Availability of Information in Public Sources

We characterize something as easy to find if similar information is available from multiple sources for multiple infrastructure targets of a similar type (e.g., all airports). Something is considered hard to find if only a single example of such information is located or no information was identified by the red team at all. Some types of information could be found for one class of infrastructure or for one scenario, but not for others.

The availability of information in public sources concerning target locations, paths to targets, and target characteristics is moderate to good. Little effort was required to find good overhead maps and imagery of most target areas; detailed navigational instructions to and from most targets were also readily available. Some structural details about a variety of airliners, cruise ships, and train cars are available. Finding blueprints and interior or subterranean maps is difficult. Also difficult to find were details sufficient to ascertain lines of sight or to support scenarios involving point-to-point fires, such as a MANPADS, or scenarios in which being observed by security forces constitutes a threat to the operation.

The availability of information about security forces, their deployments, plans, and capabilities is poor to moderate. Although the jurisdictions covering a particular geographic area and the total number of relevant public law enforcement personnel are typically matters of public record, information on specific patrol strengths, force deployments, operational plans, and security force equipment and resources is patchy and anecdotal at best. Although the red team found some relevant information in brief references contained in press releases or in composite news articles, team members found no systematic or complete listings of deployments or response plans. The team readily found police radio frequencies on the World Wide Web. Obtaining information about the infrastructure itself is easier than locating information about adjacent public and private spaces (e.g., small marinas near major seaports, residential areas, or business parks adjacent to airports that might serve as launch points for a MANPADS attack). Not only is information tagged to an infrastructure target easier to find than similar information about (often unnamed) areas adjacent to such targets, named infrastructure elements are the subject of more news articles and government reports. For example, it was easy to map and identify the height and function of most buildings on LAX proper; once outside airport boundaries, finding out about the owner, purpose, and dimensions of any given building becomes much more difficult, even though these buildings appear on easily accessible overhead imagery and some of these data are part of the public record in the form of building permits and records of real estate transactions.

The availability of information about security measures, including cameras, detectors, and scanners, and specific security countermeasures is poor to moderate. In many cases, it was possible to learn that an area has some security measure or countermeasure in place, but the red team could not, in most instances, discover the scope (where and how much of the area around the target is actively monitored) or effectiveness of these systems.[1] Information about larger, more expensive countermeasures or security vehicles is generally easier to find. That information includes, for example, the number and details of USCG cutters, details of cargo scanning lanes at a seaport, and details of airport baggage scanning systems.

The red team's ability to assess threats to operational plans is poor to moderate. While assessing security force capabilities was uniformly difficult across scenarios, this difficulty is more likely to undermine the success of terrorist attacks in some scenarios than others. Thus, the red team was better able to estimate threats to operational plans for such scenarios. Where their ability to assess such threats was diminished, it was because a particular terrorist operation and attack mode was much harder for security forces to interfere with. For example, a terrorist group executing a MANPADS attack on a passenger airline from a location that is not actually part of an airport is not likely to encounter security forces beyond local police patrols. In addition, the nature of the attack is such that it would be more difficult for security forces to interfere with the outcome once the MANPADS has launched. In contrast, a terrorist who is attempting to deliver a chemical agent to a congested subway terminal may be more likely to encounter security forces with some training in identifying suspicious behavior and response. Consequently, there is more opportunity for security forces to interfere or influence the final outcome of the attack. It was also often difficult to determine when and if terrorists might come into contact with security forces and what ability security forces may have to detect and interdict an operation in progress. However, depending on the method of the attack and the location from which it is launched relative to the target, the terrorist's limited understanding of those forces may not diminish an attack's success.

A plethora of information related to commuter train and bus security in and around New York City surfaced in public news sources on the Web following the attacks in London in July 2005. As a general rule, the more salient the potential threat, the more information appears

---

[1] For example, it is easy to discover that LA/LB has an extensive camera and surveillance network, but it is hard to get details about the locations, ranges, or coverage details of those cameras. Thus, an adversary must operate as if observed while in the area of operations. Fortunately, adversaries may not be able to ascertain whether it is possible to thwart that observation. Unfortunately, adversaries, knowing they may be observed, can plan their operation assuming that they are observed and maximize activity camouflage accordingly.

on that threat and on the security measures designed to mitigate or prevent it. More discussion in the public sphere makes more information available to terrorists who are monitoring publicly available, off-site data sources. However, the usefulness or accuracy of information sources, particular those from unedited sources such as personal blogs or Web sites, is not always apparent.

Particularly terrifying attack modes may generate attention from defenders independently of the probability that they may occur. For example, given the threat of nuclear or radiological weapons being smuggled into the United States, considerable concern has arisen about the security of cargo, which means that a good deal of information about the type of security measures and countermeasures—or lack thereof—has entered the public domain through congressional testimony, product descriptions, news accounts, and assessments of the efficacy of security measures or processes.

Stoplight Summary

Table 4.1 presents a stoplight summary of all information found by the red team for the six scenarios, presented from the defender's perspective. The stoplight chart graphically depicts the project team's evaluations of the sufficiency of the information available, in terms of whether it could impact the terrorists' prospects for a successful attack operation in each scenario (i.e., terrorists could use it to make operational decisions). The color codes in Table 4.1 reflect this assessment. While Chapter Two identifies ModIPB categories that are critical to the operation in each scenario, Table 4.1 does not weigh or otherwise distinguish the relative importance of different categories beyond asserting their possible relevance.

The assessment was conducted by the project leaders based on information learned through the previously described interviews with infrastructure owners and operators and discussions with the RAND SMEs. *The color codes do not indicate prospects for success (or failure) for a particular scenario. Nor do they indicate any type of discovered weakness or vulnerability in a category.* Rather, the color codes indicate only the sufficiency of the information found by the red team for planning the attack in each scenario. Red (R) is an indication that sufficient information was found by the red team for operational planning purposes. Yellow (Y) indicates some information was located but the level of detail was not likely sufficient for effective operational planning. Information that fits into this category is, by far, the kind that was most frequently identified in this investigation. Green (G) indicates that the red team found little or no information for that category. White (W) indicates that information in that category was deemed to be irrelevant to the scenario under consideration.

As noted in Chapter One, having information about a countermeasure may either encourage or discourage terrorist plans to strike a target. This can be illustrated with two hypothetical examples. In the first, terrorists discover that LAX-bound aircraft carry defenses against MANPADS; this information discourages the terrorists. In the second, terrorists acquire detailed architectural drawings of the Times Square station. This information encourages the terrorists, in part, by increasing the likely impact of any attack. Nevertheless, in both examples the relevant subcategory under "characteristics of the target" would be colored red because, in both cases, the terrorists acquire information that they need for their planning. The information presented in Table 4.1 assesses only how much information for each category the red team located, and does *not* indicate whether the availability of this information increases or decreases the likelihood of an attack, the probable success of a given scenario, or the likelihood of a given scenario occurring.

Table 4.1
Stoplight Scorecard of Modified IPB Categories for All Six Scenarios

Category	Rail Infrastructure		Air Transportation Infrastructure		Maritime Transportation Infrastructure	
	Scenario 1—Gas attack vs. subway	Scenario 4—Madrid-style rail attack	Scenario 2—Explosive in air cargo	Scenario 5—MANPADS	Scenario 3—Nuke in shipping container	Scenario 6—Cole against a cruise ship
1. Avenue of approach and ease of access						
Location of target	R	R	R	R	R	R
Surrounding terrain and buildings	Y	Y	R	R	R	R
Maps	Y	R	R	R	R	R
Blueprints	G	G	W	W	W	G
Types of building construction	W	Y	W	W	W	Y
"Critical points"	Y	Y	Y	Y	Y	Y
OCOKA	Y	Y	W	Y	W	Y
Available paths to target	Y	R	Y	Y	W	R
Exact path(s) to take	Y	R	Y	R	W	R
Go/no-go areas (because of barriers, obstructions or impassable terrain)	Y	Y	R	Y	W	R
Areas of restricted or limited access (security restrictions)	W	W	W	Y	W	Y
Rules and laws governing movement (vehicular or otherwise) in target area	R	R	R	R	W	R
Traffic conditions (all relevant vehicular and pedestrian modes)	Y	Y	R	R	W	G

Table 4.1—Continued

Category	Rail Infrastructure		Air Transportation Infrastructure		Maritime Transportation Infrastructure	
	Scenario 1— Gas attack vs. subway	Scenario 4— Madrid-style rail attack	Scenario 2— Explosive in air cargo	Scenario 5— MANPADS	Scenario 3— Nuke in shipping container	Scenario 6— Cole against a cruise ship
2. Characteristics of the target						
Possible locations from which to launch the attack	Y	Y	Y	Y	Y	R
Possible times or windows of time to launch the attack	Y	Y	Y	Y	W	Y
Mobility and variability of the target. If mobile, (predictable) paths it takes	Y	R	Y	R	W	R
Relevant features and structure of the target (i.e., technical details)	Y	Y	R	R	Y	Y
3a. Security forces						
Locations of headquarters, stations, checkpoints	Y	Y	Y	Y	W	Y
Overall size and types (uniformed, plainclothes, canine)	Y	Y	Y	Y	Y	Y
Number on duty at any one time, hours of duty, variation by times of day	G	G	G	G	G	G
Applicable operational jurisdictions, number of security forces contributed by each	Y	Y	Y	Y	Y	Y
Capabilities	Y	G	G	G	G	Y
Vehicles and other assets	Y	G	G	G	G	Y
On what radio frequencies are they operating?	R	R	Y	R	R	R
ROE or use-of-force policy	G	G	G	G	W	G
Specific or individual deployments	G	G	G	G	G	G
Fixed positions	G	G	G	G	Y	G
Patrols (routes, schedules, number of personnel, vehicles patrolling)	G	G	G	G	G	G
Times of observations (cameras, live operatives)	Y	G	G	Y	Y	Y

Table 4.1—Continued

Category	Rail Infrastructure		Air Transportation Infrastructure		Maritime Transportation Infrastructure	
	Scenario 1—Gas attack vs. subway	Scenario 4—Madrid-style rail attack	Scenario 2—Explosive in air cargo	Scenario 5—MANPADS	Scenario 3—Nuke in shipping container	Scenario 6—Cole against a cruise ship
Number of security personnel required to be "passed"	G	G	G	G	W	G
Variations by times of day	G	G	G	G	W	G
Security plans (operational details)	Y	G	G	G	Y	G
Security response plans	Y	G	G	G	G	G
Past performance with previous (similar?) incidents	G	Y	G	G	Y	Y
Behaviors/plans/capabilities at different levels of alert	Y	Y	G	G	G	Y
Response times	Y	G	Y	G	G	Y
3b. Security measures						
Kinds of checkpoints to be passed	Y	Y	Y	R	Y	R
Search procedures (For what will officials be looking or asking?)	Y	Y	Y	Y	Y	Y
Cameras, scanners, or detection equipment operating over the area to be traversed	Y	Y	Y	G	Y	Y
Sensitivity of these devices	G	G	Y	G	Y	G
Frequency with which sensors are "read"	G	G	Y	G	Y	G
Illumination	R	R	W	G	W	G
Specific countermeasures (e.g., vehicle barriers)	Y	G	Y	Y	Y	Y
3c. Other population groups						
Other people at the facility (Why are they there? What are they doing?)	Y	Y	G	G	Y	Y
Bystanders, recreational users, passengers	Y	Y	G	G	Y	Y

Table 4.1—Continued

Category	Rail Infrastructure		Air Transportation Infrastructure		Maritime Transportation Infrastructure	
	Scenario 1—Gas attack vs. subway	Scenario 4—Madrid-style rail attack	Scenario 2—Explosive in air cargo	Scenario 5—MANPADS	Scenario 3—Nuke in shipping container	Scenario 6—Cole against a cruise ship
Differences in population at different times of day	Y	Y	G	G	G	G
Vigilance instructions and emergency phones	R	R	G	G	G	G
Level of security training for nonsecurity employees of the facility	R	R	G	W	G	G
Schedules of regular arrivals and departures from target area	R	R	Y	R	G	Y
Ease of camouflaging yourself as a member of one of these population groups	R	R	Y	Y	G	R
4. Threats to the operation						
Threat posed by security forces and measures	Y	Y	Y	G	Y	Y
Deployments, response times, vehicles, equipment, plans	G	G	G	G	Y	Y
Cascading information (from organization of oversight and headquarters locations to who will be on the avenue of approach on attack day)	G	G	G	G	G	G
Estimated effectiveness of security response capabilities (including communications)	Y	Y	Y	G	Y	Y
Threat posed by employees of the target	Y	G	Y	W	Y	G
Citizens (concentrations of, heightened vigilance of)	Y	Y	W	G	R	G
Weather (as it affects effectiveness of the operation)	Y	W	W	R	W	R

G = little or no information discovered.

Y = some information discovered.

R = sufficient information discovered for planning purposes.

W = category irrelevant to scenario under consideration.

The six scenarios are grouped as follows. Information collected for scenarios 1 and 4 (rail infrastructure) are located in the left two columns. Information collected for scenarios 2 and 5 (air transportation infrastructure) are located in the middle two columns, and information collected for scenarios 3 and 6 (maritime transportation) are located in the two columns on the right. We have organized our findings in this way to help the reader compare the sufficiency of the information collected for scenarios that feature different kinds of attacks on similar infrastructure targets.

Because evaluations focused on the sufficiency of information for a given scenario, the same type of information might receive a different rating in the context of other scenarios in which that type of information might be of greater or lesser importance. Much of the observed variation in the evaluations depicted in Table 4.1 is specific to the scenario rather than general to a class of targets. However, a few examples of categories of information in Table 4.1 show similarities across infrastructure types. For example, information about "vigilance instructions and emergency phones" is green (G) for scenarios targeting the rail infrastructure but red (R) for the scenarios targeting the air and maritime transportation infrastructures.

In general, it was relatively easy to find information, including maps and images, that showed the physical location and the geographical region around potential infrastructure targets. However, information about security measures or specific security practices was more difficult to locate. Similarly, it was difficult for members of the red team to determine how the information that was located would impact the overall terrorist scenario.

Implications of the Availability of Information

Our analysis leads us to two observations regarding the availability of public information to both terrorists and defenders.

Despite the availability of a substantial amount of information, there is still uncertainty regarding the implication of this information on the success of a given attack. Although our results indicate that, in many cases, a great deal of information relevant to the execution of a particular terrorist attack is available, much of that information is not detailed enough to permit terrorists to execute attacks with a high level of confidence that the attack will be effective. Even when methods of ascertaining basic facts (e.g., how many police are normally assigned to a particular subway station) are relatively straightforward and the terrorist researcher can readily confirm whether this information is true, some facts critical to the success of an operation will remain unknown. For instance, even if a terrorist knows how many officers there are in a particular security force, he or she will not know how thorough security personnel will be, how they will react in an emergency, or whether their response will exacerbate or mitigate the attack's effects. For instance, after the first attack on the World Trade Center on September 11, 2001, emergency responders warned those inside the building to stay put. In that instance, relying on standard procedures in the face of an extraordinary event arguably contributed to the total number of fatalities. On the other hand, as an example of mitigation based on unexpected events and reactions to them, consider, for instance, the 40-minute delay in the departure of United Airlines flight 93, the plane that crashed in Pennsylvania on September 11, 2001 ("9/11 Recordings Chronicle Confusion," 2004). The terrorist planners could not have foreseen that delay, yet it enabled the passengers to learn from people on the ground what had happened in New York City and at the Pentagon, prompting them to rush the cockpit, which, in turn, prevented what would likely have been a deadly attack on the nation's capital.

The everyday behavior of security forces may be uncharacteristic of their actions under stress. The same is true for machines: One might determine which are installed, what their capabilities are, and how often they are running, but that does not necessarily indicate how effectively (fallible or unpredictable) security forces use them. For this reason, terrorists may have difficulty predicting how these forces will respond under most attack scenarios.

Information useful to terrorists can be obtained from nonobvious sources. The World Wide Web has expanded the amount of information available to everyone—including, of course, terrorists. However, by far the bulk of the information identified in our red-team exercise originally appeared in other, formal sources (albeit archived via the Web). Official and company Web sites provided useful information, but not to the point of constituting a threat to operational security. For example, manufacturers of equipment to detect explosives provided product specifications that included minimum resolution and penetration. News articles released information about the size of police forces, as well as information about possible suspicious behavior to watch out for. However, information such as this does not constitute a threat to operational security—that is, just because someone knows that the information exists about some countermeasure does not make the countermeasure any less effective.

The bulk of information that could be construed as a threat to operational security was found on "unofficial" Web pages put up by private citizens, Web communities, or small fringe organizations. An interesting but clearly minor fraction of the information came from discussion groups. Relatively little information was obtained through blogs. Whether that small ratio is testimony to the primitive state of blog search tools (see Stephen Baker, 2005) or simply the fact that journalists who publish in traditional venues are apt to be more focused on terrorism-related information than bloggers remains to be seen.

Policy Recommendations

Based on the findings described above, we propose two recommendations intended to help infrastructure owners increase security.

- **To prevent security details from entering the public domain, periodically review and revise procedures for operational and information security.**

Our findings indicate that information pertaining to certain ModIPB categories—for example, security force deployments (routes, schedules, number of personnel, vehicles patrolling)—is not easily accessible through off-site, public information sources. During the course of the interviews, infrastructure owners and operators either described or referred to procedures in place for sharing sensitive or security-related information. Our findings suggest that these procedures effectively prevent this information from being easily located in public sources.

Nonetheless, our red team did identify a wide variety of kinds of information, including overhead images, schematics of sites and equipment, and news reports in many different kinds of publications. Moreover, new information is being added to the public domain every day, along with new capabilities for searching and fusing information. Perhaps even more important, new *kinds* of information and methods for sharing information are becoming available; as we noted previously, the availability of videos on the Web has increased dramatically, and this trend is likely to continue. Other kinds of information not yet imagined, as well as new search mechanisms, will likely emerge over time.

Thus, procedures for securing sensitive information should be evaluated regularly, taking into account developments in technologies for storing and retrieving data, with a view toward identifying vulnerabilities that might allow sensitive information to enter the public domain. Elements of an evaluation process may include (but are not limited to) comparisons of both human and technical procedures for protecting information systems—and related documents—that concern the activities of security forces, expert review of computer systems in which information about the character and activities of security force procedures and other sensitive information is stored, and auditing external Web sites and information sources to ensure that they do not disclose information that should be protected.

- **Include information that can be obtained from easily accessible, off-site public information sources in vulnerability assessments.**

Our study identified a great deal of information that was easily accessible from off-site, public information sources that terrorists could potentially use to plan attacks. In most cases, this information is necessarily in the public domain because it provides an important service. For instance, as noted above, ecommerce sites for organizations that provide passenger travel must present information about schedules and costs, as well as information about security procedures that passengers need to determine what they can take with them and to plan departure times. Rather than try to eliminate this information from the public domain, owners and operators of transportation businesses must instead assess the specific vulnerabilities created by public knowledge of such procedures and address them through countermeasures.

Given the variation in publicly available information across attack scenarios and kinds of transportation infrastructure identified in this report, owners and operators of particular kinds of transportation infrastructures will need to determine how information available in the public domain is likely to affect *their specific business or organization's* vulnerability. In summary, vulnerability assessments or risk analyses should take into account what information can be acquired easily from off-site, public information sources, both to improve knowledge of a business's own specific vulnerabilities and to adjust security procedures if necessary.

Finally, owners and operators of transportation infrastructure organizations must determine how frequently vulnerability assessments should be conducted to ensure that, as new information enters the public domain, in those assessments capture it. During our interviews with them, infrastructure owners and operators described varying levels of effort—ranging from general familiarity with news and media reports to hiring specialized contractors and vulnerability assessment experts—to monitor or assess the amount of public-source information available. Given the large number of media outlets and information dissemination channels available today, firm-specific information that might be relevant to the security of organizational assets and operations could crop up in publicly available sources at any time. We cannot specify a priori how frequently such reviews should be conducted, but we believe they should either be integrated into current vulnerability assessments or, if conducted separately, should be carried out with at least the same frequency.

In addition, no evidence from this investigation suggests that removing information from the Web, or from the public domain more generally, would alter the risk of a given scenario occurring.

Summary

In this report, we have introduced and established the utility of the ModIPB framework, which provides a set of categories that can be used to guide analyses of publicly available information concerning transportation infrastructures that might be useful to terrorist planners. Given the enormous amounts and diverse kinds of possibly relevant information currently available and the likely emergence of new forms of information in the years ahead, such a framework is needed to direct searches of information in the public domain to ensure that all relevant topics are taken into account. The results of a red-team exercise in which team members relied on this framework to gather information pertinent to each of six scenarios—two each in air, rail, and sea transportation infrastructures—demonstrated its value in that an expert review confirmed the comprehensiveness of the information that the red team collected.

The red-team exercise results also call attention to the existence of both large amounts and diverse kinds of information relevant to planning terrorist acts in the public domain. Given the possibility that information present in these archives might reveal vulnerabilities of interest to terrorists, we recommend that owners and operators of transportation infrastructures regularly examine public information sources, relying on the ModIPB or a similar framework, for guidance in determining whether information currently in or entering the public domain constitutes a threat to the security of their assets and operations.

Finally, the evidence presented here regarding the availability of information relevant to elements of the U.S. transportation infrastructure confirms the importance of including such sources in organizational vulnerability assessments.

What the Red Team Found

In the red-team simulation, RAND's red team focused on gathering information in a way that involves zero-to-low risk and requires only minimal expertise. One method fitting this description is searching the Internet via search engines. Although this type of information-gathering may happen at any point in the planning process, its logical position is early. For each scenario, we assume that the target and mode of attack have been selected but that no other knowledge has yet been collected.

Three main information needs at this point in operational planning could result in showstoppers if not satisfied.

First, terrorist planners need to know whether countermeasures exist against the attack mode that they are planning and whether such countermeasures are in place at the planned target. For example, terrorists planning a MANPADS attack against a passenger aircraft need to know whether there is a technology that thwarts MANPADS attacks (e.g., flares against infrared missiles). Then, they need to know whether the target might be equipped with them (at present, airlines in the United States are not).[1] If so, details of the parameters, limits, and counter-countermeasures for the countermeasure will be required.

Second, terrorist planners must have a sense of the extent of their exposure and vulnerability to security forces. Many variables affect exposure to security. One is the consequence of detection. Certain operations will fail if detected at any point during a prolonged execution. Consider scenario 3, the transshipment of a nuclear device in a cargo-shipping container; if the nuclear device is detected and discovered at any point during shipment, the operation is considered a failure.[2] Another variable is the duration of exposure. For a USS *Cole*–style attack, that duration is the time it takes to traverse the last 100 yards to the ship. For a MANPADS attack, exposure is limited to the time the weapon is taken out of concealment and shouldered. The longer an operation is exposed to possible countermeasures, the more information about these countermeasures will be required. Also important is the control of

---

[1] According to Stoller (2005), such countermeasures are presently in the testing phase, and not currently employed. On the other hand, Vause (2004) indicates that El Al began equipping their planes with an antimissile system called Flight Guard in May 2004. Also see Chow et al. (2005).

[2] If terrorists merely wished to detonate the device in the port prior to inspection—which would still be deadly and costly—their information needs would be different. Nevertheless, in order for the red-team members to focus their information-gathering activities, each scenario had to have specific information requirements. Therefore, for the purposes of this red-team exercise, any outcome other than the scenario's stated objective was considered a failure.

both the transit and attack modes. At one extreme, the attacker controls neither the transit mode nor the attack mode, such as in scenario 2 in which a weapon is dropped off at a shipping counter with the hope that ground crews will put it into the cargo hold of a passenger airline. At the opposite end of the spectrum is an attack in which the terrorist controls both transit and weapon, such as a MANPADS or suicide vehicle attack. In the middle are attacks in which the terrorist has control over the weapon but does not have control over the transit mode, such as using a train or bus as a conveyance to the attack site. The less control the terrorists have, the more extensive their knowledge of security must be.

Third, planners need to obtain certain information about the target. What and how much will vary depending on the proposed attack scenario. To launch an attack against a cruise ship, it may be that the only information required about the target is where it will be physically located at points in time. To maximize the effect of such an attack, structural details about the ship may be necessary. Similarly, simply releasing sarin gas in a subway may require very little information about the target station; to maximize such an attack requires information about the layout, the ventilation system, train schedules, congestion levels, and some knowledge of emergency response plans.

Red-Team Procedure

The red team, made up of RAND research assistants, served as proxies for terrorist researchers. The red-team members were consistent with the criteria described in the al Qaeda manual, which stated that information gatherers should be college educated, with access to broadcast and print media. The red-team members had no specific background in the infrastructures that they were investigating. The instructions for the information collection phase are as follows:

- Each team member was initially assigned two classes of transportation infrastructure target to investigate: either the air and rail infrastructures, the air and maritime transportation infrastructures, or the rail and maritime transportation infrastructures. At this phase of the investigation, specific targets or modes of attack were not assigned and each red-team member was instructed to build a familiarity with the particular infrastructure and collect any information about possible vulnerabilities or security measures. This information along with public documents that the project leaders collected on possible terrorist attacks, identified in Chapter One, were used to develop the six scenarios.
- During the next phase of the investigation, each team member was assigned two scenarios, one for each class of infrastructure that they had investigated from the previous phase that included a specific target and mode of attack. The red-team members were also provided with the ModIPB framework listed in Table 4.1 and instructed to search for data for each relevant category. All information-gathering activities were completed before August 2005.
- At the end of the information-gathering phase, RAND SMEs reviewed the collected information and compared it with information collected by the project leaders from infrastructure owners and operators during interviews at the early stage of the study.

- A subset of the information collected from three of the six scenarios, one for each infrastructure type, was compared to that collected by a senior RAND expert with more extensive experience in red-team information-gathering activities.

The discussion for each of the six scenarios first provides the mission's objective and details and then identifies the information items in the three critical areas—drawn from the ModIPB—that operatives would need to start planning that mission, followed by the results of the red-team information search. The three scenarios that, on their faces, appear to have the highest information-gathering requirements are presented first; those that have lower requirements follow.

A limitation of our approach, and indeed of any effort to plan a terrorist operation, is that it is not always possible to specify ahead of time what information items will prove to be critical. This is especially true with regard to "showstoppers." Critical information requirements were roughly identified during the development of the ModIPB framework in conjunction with the six identified scenarios. Refinement of the critical information elements occurred during the red-team information-gathering activity and subsequent analysis of the information gathered for all of the ModIPB information elements.

Scenario 1: A Poison Gas Attack (NYC Subway)

In this scenario, a terrorist group plans to release a chemical agent (for example, sarin) at the 42nd Street/Times Square Station (one of the busiest stations of the New York City subway system [New York City Subway, 2002]) during the afternoon rush hour. The terrorists hope to spread the poison gas through the entire subway station. The plan calls for five terrorists to arrive, individually, at one of five different entrances to the station, coordinating their schedules through cell phone text messaging. Each terrorist carries a small piece of luggage (a suitcase or backpack) modified to release an aerosolized chemical agent. When all five are at their appointed locations, each will proceed to a predesignated platform; once there, they will set down their weapons, activate timers that release the poison gas in 10 seconds, and depart.[3] This scenario is reminiscent of Aum Shinrikyo's March 1995 attack on the Japanese subway system that killed 12 and injured thousands (Murakami, 2001).

Critical Information for the Mission When No Other Information Has Yet Been Collected

- Security measures: What security monitoring devices and sensors, including chemical or biological sensors and other monitoring equipment (e.g., cameras), are present in the station? Such sensors determine whether a device will be detected before it dispenses gas or before it dispenses an effective amount of gas. It also includes information about what types of countermeasures are present in and around the station.

---

[3] For purposes of discussion, we view this as an optional feature of the scenario. The terrorists can be expected to research exit routes and modalities, but should they fail to satisfy themselves on that score, they are nevertheless expected to carry out the attack. Conversely, insufficient confidence that the operation can succeed would abort further mission planning.

- Security forces: What is the exposure to security forces and what security presence and procedures are in place? This information is necessary to determine the odds that the terrorists will be stopped before setting off the devices.
- Avenue of approach: Specific location of entrances, exits, and platforms layouts will help in determining what obstacles may be encountered and where to release the gas.

Security Measures

Sensors. News sources were among the most effective information sources with respect to security measures on the New York City subway. In 2004, *The New York Times* described how the New York subway would acquire "shoebox-size sensors that analyze the air and would sound a silent alarm in the event terrorists unleash a biochemical attack" (Weiss and Lisi, 2004). The article also detailed the testing of a biological sensor that was intended to monitor for the presence of agents that may be used in a biological attack. A March 2005 article in *Newsday* noted that Grand Central Station was the only station in the New York subway system that possessed an electronic chemical agent sensor in the main terminal (Sanchez, 2005). In a July 2005 article in the *New York Sun,* the MTA said that it was investigating sensors that could detect anthrax and sarin but such sensors were only in the testing phase; deployment schedules and costs had yet to be determined by the MTA (Smerd, 2005). According to the article, all of the devices that were located in Grand Central Station and Penn Station were being tested for efficacy. The article suggested that little else was currently being done to detect chemical or biological agents.

Investigations into the type of surveillance in and around the target revealed an article that discussed MTA's intent to expand the use of closed circuit cameras throughout the subways (Luczak, 2005). Yet, according to Trager (2005), although "the MTA has used some of the security money to install more surveillance mechanisms—such as closed circuit cameras—throughout the subway system, it has also dropped its staff, who monitor the cameras, by 10 percent," so less actual monitoring may be taking place. The article also mentioned that the New York City subway system was also accepting proposals for technologies that could prevent a cell phone from being used to trigger a bomb. The MTA reportedly spent 5 percent of the $600 million it committed to improving transit security, and most of that was focused on public awareness (Chan, 2005).

A March 2005 *Newsday* article contained information on a test carried out in 1966 during which the Central Intelligence Agency and U.S. Army spread harmless biological agents to study how they would be dispersed through the system.

> Testers broke light bulbs on gratings on the Seventh Avenue and Eighth Avenue lines. The bulbs contained a harmless anthrax cousin, *Bacillus substilis* variant *niger.* . . . Test results show that a large portion of the working population in downtown New York City (New York City) would be exposed to disease if one or more pathogenic agents were disseminated covertly in several subway lines," the team reported. . . . The air currents generated by moving trains disseminated the spores. An estimated 1 million people were exposed. (Sanchez, 2005)

Although a 1998 article pointed out that the only active air handling system was at Grand Central Station,[4] a more recent article in *Engineering News Record* observed that Times Square Station would be equipped with a new ventilation system as part of Phase One of a three-phase station renovation (Cho, 2004). According to a March 2004 *New York Times* article, Phase One of the renovation was completed in September 2002 (Dunlap, 2004). The red team discovered nothing to suggest that ventilation systems capable of mitigating gas attacks existed at Times Square Station. The potential over-platform and under-exhaust systems, along with air flow from moving trains, may help to distribute the gas around stations, as suggested in the 1966 biological terrorism experiment.[5]

Several sites discussed the NYC subway's "Eyes and Ears Program." One article noted that employees are trained to watch out for suspicious behavior such as a passenger's wearing of a winter coat in summer (Robin, 2005). Yet, several employees questioned its usefulness noting that the "training" consisted of a 15-minute video and some pamphlets discussing terrorism. It added that, following the attacks on the London underground on July 7, 2005, transit officials hired a contractor to evaluate the transit employees. The MTA official said that the report from the contractor suggested the MTA transit employees were "adequately" on guard (Robin, 2005).

An April 2005 Web article discussed the *absence* of countermeasures, noting that, in event of a terrorist attack, there were gas masks for the transit workers only (Trager, 2005).

Security Forces

Terrorists may need a sense of how many security officers they might encounter throughout the subway system under normal conditions—something they might figure out by knowing how many police personnel are at work at any one time. In March 2004 (just after the Madrid bombing), *The New York Post* reported "2,838 [New York City Police Department Transit Bureau] cops . . . work for 12 subway districts" (Weiss and Lisi, 2004). In addition, 692 MTA officers patrolled Metro-North and Long Island Rail Road stations; furthermore, the MTA planned to bolster its police force from 521 officers in 2001 to 723 by the end of 2004 (Weiss and Lisi, 2004). The article added that subway security occurs 24 hours a day and 7 days a week, with police on 12-hour shifts (City of New York Police Department, undated).

After the July 2005 bombing in the London underground, WABC reported that the New York City Police Department was "doubling the 2,700 officers assigned to the Transit Bureau" (WABC, 2005), and that, during the week of July 10, every rush-hour train would have at least one officer (some in plainclothes). Police officers currently conduct surprise security sweeps on subway cars before they enter a tunnel or cross a bridge. Such sweeps, involving up to a dozen police officers, typically last a few minutes and focus on spotting suspicious items (Weiss and Lisi, 2004).

---

[4] "The city's subway system, according to [geologist Dr. Andreas] Pflitsch, serves about 3 million people daily, and, except for Grand Central station, all other subway stations are equipped with only passive air ventilation, like entrances and exits to and from the subway, stairs between levels and subway ventilation located on the city sidewalks." Pardo (1998)

[5] Sanchez (2005). Note, incidentally, that, although ventilation systems may help disperse a chemical attack and thereby limit casualties, it would also disperse biological agents and could thereby increase casualties in such a scenario.

According to a May 2004 *New York Times* article, the New York Police Department provided all officers with a pocket-sized reference card containing information on effectively spotting potential terrorist activities (Rashbaum, 2004). A July 4, 2004, *New York Post* article reported,

> The department tells its officers to watch out for people with drivers' licenses from more than one state, passports from more than one country, and identification papers with different names and people videotaping or photographing bridges, tunnels, utilities, landmarks and government facilities. (Guart, 2004)

It also advises officers to watch for anyone who is "overtly hostile" and expresses "hatred for America and advocates violence against America and/or Americans."

In practice, however, there may be gaps in communications among emergency responders, metro transit officials, and commuters in case of an emergency. An August 2004 blog entry reported that a small fire on the track of an underground train created panic on the train when the conductor proceeded to extinguish the fire without informing passengers (Mitchell, 2004). The presence of smoke, in the absence of information, prompted several passengers to kick out train windows to exit the train.

Information on police radio frequencies and police ten-codes[6] is readily available (see Table A.1 for frequencies). One resource provided information on more than 400 different frequencies used by police, emergency responders, LIRR, and even Amtrak and others throughout New York City, many updated and monitored up to once every 24 hours (RadioReference.com, undated).

Table A.1
New York City Police Frequencies

Frequency	Agency	Description
160.9050	NYC Transit Police Department (NYCTPD)	Div 1—Manhattan South
161.1750	NYCTPD	TAC 1—Manhattan South
160.6950	NYCTPD	Citywide
161.1900	NYC Transit Authority (NYCTA)	Interborough Rapid Transit subway
161.5050	NYCTA	Brooklyn-Manhattan Transit subway
160.8450	NYCTA	Independent Subway

SOURCE: RadioReference.com (undated).

---

[6] Ten-code is a shorthand way of describing police orders by using numbers. For example, "10-4" is an acknowledgment, while a "10-31" refers to an explosive device or threat (n2nov.net, 2005).

Avenue of Approach

As noted above, the Times Square Station is undergoing a multiyear renovation. Diagrams of the station platforms were available from two different architectural firms. The first set were part of a 1996 study of the ventilation options for Times Square Station, and the second set were available from one of the architecture firms that was working on the Times Square project. These are shown in Figure A.1. The diagrams presented in Figure A.1 can help terrorists in identifying optimal points of release for poison gas. Phase One of the multiyear project was completed in 2001 ("42nd Street," undated) and included the planned improvements to the existing ventilation system (Cho, 2004).

What the Expert Searcher Found

A public document, National Fire Protection Association (NFPA) 130, *Standard for Fixed Guideway Transit and Passenger Rail Systems* sets forth standards for emergency ventilation systems: For example, fans should be able to achieve full operating speed in 60 seconds. It also covers emergency egress and other safety factors associated with fixed-rail transit systems (National Fire Protection Association and American National Standards Institute, 2003). The information contained in the standards documents provides specifications for the removal of smoke from the platform in case of fire. This would provide some information to terrorists about what type of ventilation could potentially be used in the case of an emergency. However, the NFPA document does not discuss any specifications for ventilation in the case of the release of a chemical or biological hazard in the station. It is also unknown whether MTA employees would use the emergency ventilation system in the case of a chemical or biological gas attack.

Summary Findings for Scenario 1

The publicly available information that was collected was relevant for addressing three questions: (1) what were the odds that a terrorist would encounter a security official on the way to the platform? (2) what sensing devices were available to monitor the platform? and (3) how would the ventilation system affect chemical dispersion? In each case, available public information collected by the red team provided only limited information for answering the questions posed above. Furthermore, it is believed that any additional public information would not prove helpful. Additional information from tools such as sophisticated models of commuter traffic flows and air circulation might be useful, but the ability to use such tools are not necessarily within the capabilities of most terrorists. No further information on countermeasures was gathered, which suggests but does not prove that no such countermeasures existed. The July 7, 2005, bombings of the London transit system have raised security consciousness throughout the NYC subway system and may alter the state of security plans and operations; much depends on how long the state of heightened security lasts.

Figure A.1
Schematic Diagrams of Times Square Station

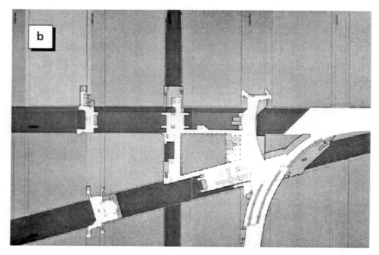

a

Lower
mezzanine

Upper
mezzanine

7th Avenue
Line (IRT)

Shuttle Line
Terminal
(IRT)

Flushing Line
Terminal
(IRT)

Stairs to mezzanine

Broadway Line
(BMT)

Island-type platform

©Metropolitan Transportation Authority. Used with permission.

b

©Metropolitan Transportation Authority. Used with permission.
SOURCES: Upper figure: Li and O'Dwyer (1996). Lower figure: William
Nicholas Bodouva and Associates (undated).
NOTES: The upper figure comes from a 1996 ventilation study; the lower
figure is from an architectural firm currently involved in renovation of
the station.

RAND *TR360-A.1*

Scenario 2: Bomb in a Passenger Plane Cargo Hold (at LAX)

It bears noting that almost a third of all domestic (and almost half of all international) air cargo travels on passenger aircraft. There are 50 carriers and 2 million shippers in the market (Elias, 2005). Many major commercial passenger airlines also have a shipping service for shipping packages and provide an important source of revenue for the airlines. For most airports, the package shipping counters are not located in the same terminal as the passenger counters and therefore have different screening processes.

In this scenario, terrorists ship an explosive device via the passenger airline's cargo shipping service. The device's power is sufficient to cause the aircraft to crash and the device is designed with a timer that is set to detonate once the aircraft is airborne. The terrorists also hope to instigate further economic losses on the U.S. airline industry. The precise time at which the package is dropped off at the airline cargo-shipping desk is designed to maximize the odds that it will be shipped on a passenger airline (as opposed to an all-cargo flight), typically one hour before the designated flight on which one wishes the package to travel. In preparation, terrorists first attempt to register as a "known shipper" and ship a full-sized explosive device. Failing that, terrorists try to ship one or more lightweight devices to exploit nationwide TSA regulations that allow carriers to load packages from unregistered shippers on passenger airlines if they weigh less than 16 ounces.[7]

Critical Information for the Mission When No Other Information Has Yet Been Collected

- Security measures: What security monitoring devices and sensors are used to scan cargo before it is placed on a plane? What countermeasures are in place on the plane to protect it from explosions? Security procedures used for screening packages and procedures associated with the Known Shipper Program (Transportation Security Administration, 2003).
- Security forces: What types of security presence are the terrorists likely to encounter? To what security procedures are the package likely to be subjected before shipment?
- Target: This includes information about the target such as the presence or absence of onboard blast resistance of some kind.

Security Measures

In 2004, Congress asked DHS to "research, develop, and procure certified systems to inspect and screen air cargo on passenger aircraft at the earliest date possible" (Public Law 108-90) Pending the availability of such technology, DHS was charged with enhancing the known shipper program (Public Law 108-90). A search of the Web suggests that screening for explosives in air cargo, for either passenger or cargo-only flights, is still in the experimental stage. A 30-day test at Boston's Logan Airport in 2003 examined the feasibility of scanning 100 percent

---

[7] Investigators believe that the bomb that took down Pan Am 103 over Lockerbie, Scotland, in December 1988 weighed less than 16 ounces.

of air cargo going onto passenger planes.[8] In 2004, TSA spent $26 million to test and evaluate explosive detection systems (EDSs) at several airports (Wilson, undated). Industry sources added that EDS were used in air cargo screening pilot programs lasting 30–90 days in the "Los Angeles, Chicago and Miami airports."[9] Of note is that some of the characteristics of such systems were available on publicly accessible vendor Web sites or through secondary sources. For example, L3 Communications mentions the specific power of their x-ray source for the VIS108 Automated Explosive Detection Systems (75 kilovolt pulsed/150 kilovolt pulsed source), the minimum resolution (38 American Wire Gauge, or 0.1016 mm diameter tinned copper wire), and the penetration of the source (30 mms of steel).[10]

Our red team was unable to find the actual percentage of passenger aircraft–carried cargo that is inspected, as this information is not easily accessible in open sources. TSA, for its part, is unwilling to release the exact percentage of cargo that is shipped on passenger planes, citing security reasons (Carrillo, 2005). But note these findings:

- A 2003 Congressional Research Service report on Air Cargo Security noted that only "5% of cargo placed on passenger airplanes is screened" (Elias, 2003), a figure repeated in 2004 by the Center for American Progress.[11]
- Although DHS claims that the regulatory minimum for inspecting cargo on passenger planes is 10 percent and some carriers may exceed this threshold (DHS, FY2006), a congressional report stated that the TSA had not yet implemented the legislation intended to triple the percentage of screened cargo carried on passenger airplanes.[12]

A 2003 newsletter from alliedpilots.org cites comments from Captain Jay Norelius from the Allied Pilots National Security Committee, suggesting that the "unknown shippers," who can ship items up to 16 ounces that can be shipped on passenger carrying aircraft, are not required to open their packages; they need only document the contents on paper (Pauwels, 2003).

A 2002 *Washington Post* article reported numerous security loopholes in cargo security procedures at the time that a draft report by the inspector general of the Department of Trans-

---

[8] L-3 Communications Security and Detection Systems, a leading developer of x-ray screening technologies, conducted the test. See Massport (2003).

[9] The prime contractor was L-3 Communications Security and Detection Systems. The program examined different configurations of equipment to test the feasibility of such a system. See Wilson (undated).

[10] Many vendors that at one time provided downloadable technical specifications for their detection equipment have removed that information from their external Web sites and will only provide it upon customer request. However, many resellers or distributors of scanner technologies will make the vendor specification sheets available for download. For example, Bavak Beveiligingsgroep BV in the Netherlands provides links to specification sheets for many of L3's detection systems (Bavak, undated; L3 Communications Security and Detection Systems, 2002).

[11] The same statement reported that TSA has ruled out screening 100 percent of cargo because of the potential negative economic impact on the airline industries (Schumer, 2004b).

[12] See U.S House of Representatives (2005) and Coalition of Journalists for Open Government (undated) for a discussion.

portation (DoT) and a TSA investigation identified (Schneider, 2002). They included the following:

- Some packages from known shippers are subject only to exterior inspection.
- Terrorists can acquire a known shipper's registration number, which is not considered classified information.
- Terrorists could easily become a government-recognized freight forwarder.[13]
- The requester of a cargo shipment could also be considered the shipper. This could allow an individual or organization without known-shipper status to avoid additional security procedures by listing the recipient as the shipper.

Many of the 50 airlines that transport cargo on passenger flights advertise this service on their Web sites. Continental Airline's QUICKPAK® service, for instance, guarantees that a package can be carried on the next available (domestic) flight, provided that it arrives at least 30 minutes before its departure (see Continental Airlines Cargo, undated[a], undated[b]). Under to the Delta Airlines Delta Dash™ program, individuals on the known-shipper list can tender items up to two hours before flight departure. Other airlines offer competing services. Using an airline's cargo and passenger Web portal services, it is possible to determine which passenger flights carry cargo (Northwest Airlines, undated[a]; Continental Airlines Cargo, undated[a]; US Airways, undated; Delta Airlines, undated). For example, using an airline's cargo-handling Web portal service, it is possible to identify flight numbers of planes that will carry a piece of cargo depending on when and where the item is dropped off and the final destination. Those same flight numbers can be cross referenced, and checked against departure and arrival information for the airline's passenger carriers.

Another way a terrorist organization might attempt to place an explosive device on a passenger plane is by registering directly with the known-shipper program. The known-shipper program enlists frequent users of the air cargo system in ensuring their freight's security, allowing security resources to be directed to nonmembers' freight. A purpose of the known-shipper program is to prohibit high-risk cargo from being transported on passenger aircraft (Security Management Online, undated). According to a May 2004 interview, TSA representatives stated that hundreds of thousands of companies were listed in the known-shipper database (I-Team, 2004). To become a known shipper, individuals or businesses must register with the airline they wish to use and submit to a search of their business premises by the airline (Calhoun, 2005). But according to a 2005 GAO report, airlines that collect information as part of the Known Shipper Program are currently not required to submit information from their known shippers to TSA. According to TSA, one of the purposes of a centralized database is to establish a method, based on information regarding the shipper, to conduct a risk assessment for each package shipped through the air cargo system. This system for risk assessments will be known as the Freight Assessment System (FAS) (Berrick, 2005). According to a TSA-Aviation Security Advisory Committee Freight Assessment System Working Group presentation from

---

[13] Freight forwarders consolidate cargo from many shippers and deliver it to air carriers. According to a 2002 GAO report, freight forwarders operate about 10,000 sites nationwide. See U.S. General Accounting Office (2002).

April 28, 2005, the FAS is still under development, with no details about how the risk of cargo is to be assessed (Transportation Security Administration, 2004b). The presentation also mentions developing a green- and red-light system for all cargo to be flown on passenger aircraft. Information about the shipper and about the cargo will be used to assign a level of risk to the specific piece of cargo. Cargo designated as high risk will require additional screening, which may include physical inspection of the contents by hand.

In addition, according to a 2004 congressional hearing quoted in the GAO report, only about one-third of all shippers are listed in the TSA's centralized known-shipper database (see Berrick, 2005). Although terrorists may not know exactly what background investigation they may have to pass to become registered, a hint of its rigor comes from a 2003 trade press report of someone who shipped *himself* through the air cargo system by completing the procedures in the known-shipper program (Coalition of Airline Pilots Associations, 2003).

The red-team investigation also found that the process for becoming a known shipper varies among airlines. Northwest Airlines (undated[b]) requires individuals to request information on how to become a known shipper. Delta Airlines (undated) provides an online registration form. Terrorists can also learn a lot from Web site discussion boards. According to information that can be found on one site (Harford Reptile Breeding Center, undated),

- Delta Airline's Delta Dash™ recommends using a small package for shipping snakes for known shippers.
- According to TSA regulations, animals are not placed through x-ray machines (Transportation Security Administration, undated).
- Boxes with snakes in them shipped through Delta Dash™ were, at most, visually inspected to ensure that there was, in fact, a snake in them (Button, 2005).

This same discussion group included the information that the only security measures faced by a snake-owner attempting to establish himself as a known shipper was an on-site interview, conducted by Delta airlines personnel whose only concern was to confirm that he was, in fact, shipping snakes. The information that was collected suggested that it would not be very difficult to become certified as a known shipper with a particular airline. This is consistent with findings of a Congressional Research Service report that stated that

> critics of existing known shipper programs argue that currently very little investigation of known shippers is required to demonstrate that these shippers are trustworthy and have adequate security measures in place to ensure the integrity of their shipments. (Elias, 2005, p. CRS-12)

TSA announced in April 2004 that it was beginning an effort to use current explosive detection technology to screen noncontainerized "break-bulk" cargo (Transportation Security

Administration, 2004a), a category that accounts for about 70 percent of cargo on commercial airlines (Gibson, 2004). As far as the red team was able to ascertain, the airline does not regularly scan upon arrival cargo that a freight forwarder or other shipper has already loaded into a container or onto a pallet.[14]

Security Forces

A 2004 article in *Homeland Security* noted that airlines are stepping up efforts to inspect cargo on passenger airlines by

- randomly checking belly cargo on certain flights
- hiring additional personnel and using bomb-sniffing dogs
- evaluating commercially available devices that detect explosives (see Moorman, 2004).

The article also noted that, although no air carriers would discuss security inspection procedures, many stated that they were improving information technology (IT) systems to enhance their known-shipper programs. Several expected to have baseline systems in place by 2005, the details of which, however, are not available from the Web.

The Intelligence Reform Act of 2004

> mandates the introduction of a pilot program that would evaluate the use of blast resistant containers for cargo and baggage on passenger aircraft. Air carriers can volunteer to participate in the program, which includes financial assistance to the airlines. The TSA is required to complete within 8 months its rulemaking to amend transportation security regulations to improve the security of air cargo that is transported in both passenger and cargo-specific aircraft. (U.S. House of Representatives, 2004a)

Information that the red team collected regarding the blast resistance of today's air cargo containers was not definitive. A 1998 Federal Aviation Administration (FAA) report found that luggage containers had very little inherent blast-resistance capability (U.S. Department of Transportation, 1998). Congressional testimony by TSA Administrator David Stone in August 2004 mentioned that TSA was trying to expand the use of blast-resistant cargo liners in passenger aircraft (Stone, 2004). There are commercially available hardened cargo boxes for airline use that have some blast-resistant capabilities (Ogando, 2002; Telair International, 2002). A December 2001 report from the UK Civil Aviation Administration nevertheless listed some limitations of current blast-hardening techniques: For example, 75 percent of the aircraft operating within European airspace cannot employ cargo containers because the aircraft are too narrow, and loose luggage, if accelerated (by an explosion), may penetrate the aircraft hull, causing critical structural damage (Civil Aviation Authority, 2001).

---

[14] According to a January 2005 CRS report, pulsed fast neutron analysis (PFNA) is being considered for screening containerized air cargo (see Elias, 2005). A second report suggests that the technology developed for sea cargo container scanning may also be useful. See Moorman (2004). However, there is no information on what (if any) screening of containerized air cargo is currently being done.

What the Expert Searcher Found

Although this scenario focused on passenger planes as cargo haulers, cargo planes rely on similar technology and processes to guarantee security. At the March 2004 National Transportation Safety Board (NTSB) Air Cargo Safety Forum, a presentation by FedEx described many of the limitations of current chemical and biological detection systems (FedEx Express, 2004). The presentation listed several examples of the reasons that FedEx was not adopting some cargo screening technologies including the absence of one-size-fits-all technology solutions, problems with device sensitivity, accuracy ("false positives" and "false negatives"), real-time alarming but not real-time notification, and problems with training and ease of use of many devices. FedEx also mentioned that these technologies are often expensive and may have significant operational impact, presenting barriers to adoption by passenger and cargo airlines.

Summary Findings for Scenario 2

Public information suggests that the risks of detection for shipping explosives aboard air carriers are not high. Because variation in package treatment seems to have a strong random component, it is unclear exactly what further information would allow terrorists to make a better assessment of the actual probability of a device being detected under any given circumstances. In addition, because security requirements apply to all airlines but details about the security procedures and technologies for specific airlines could not be identified, it was not possible to determine a particular airline or cargo service that would be more susceptible to this type of scenario than another.

Scenario 3:
Shipping a Nuclear Device in a Cargo Container Through LA/LB

In this scenario, terrorists attempt to import a nuclear device into the United States through LA/LB in an oceangoing cargo container. It is intended that the device detonate at a destination outside the port.

Critical Information for the Mission When No Other Information Has Yet Been Collected

- Security measures: Security monitoring devices and sensors include radiological or nuclear detectors (e.g., x-ray, gamma ray, container inventorying, handheld device checks) located at either the port of departure or the port of arrival.
- Security forces: Security procedures include the types of security forces located at the port and information regarding procedures for inspecting cargo containers (e.g., how containers are selected for a more thorough search) as well as information about specific vulnerabilities at foreign ports that will help in selecting which foreign port will be used to ship the device. Again, if the device is detected, the operation fails.
- Target: Since this is a transshipment scenario, it lacks a "target" per se; although a specific port is targeted for transshipment, the only relevant details concern countermeasures

and security exposure. (This scenario does not follow the device to its ultimate target destination.)

Security Measures

According to a 2004 *CalTrade Report* article, LA/LB currently has only four gamma-ray scanners and one x-ray machine to inspect the 3 million containers that flow through LA/LB annually. Dockworkers believe there are too few machines (CalTrade Report, 2004). According to the report, the machines operating full time could scan 10 percent of all incoming cargo or about 280,000 containers per year. As of April 2005, inspection devices at LA/LB consisted of five full-truck gamma-ray and two x-ray scanners, along with personal radiation detectors and radiation isotope identifier devices (RIIDs)—handheld devices that combine real-time, independent radiation sensors.[15]

According to a June 2005 *Journal of Commerce* article, by the end of 2005, LA/LB is scheduled to have radiation portal monitor[16] (RPM) coverage for all incoming cargo containers ("Nuke Detectors for LA-LB," 2005). As the Secretary of Homeland Security has reported,

> Initially, three major container terminals at the Port of Los Angeles will have their RPM systems on-line by the end of this month [June 2005]—the . . .Trans Pacific Container Service Corp. facility; the . . . Pier 300 terminal; and the giant . . . Pier 400 complex. ("Nuke Detectors for LA-LB," 2005)

A total of 90 RPMs, which will screen all inbound international container traffic and vehicles leaving the LA/LB facility for nuclear materials or hidden sources of radiation, is planned to be operational by December 2005.

Terrorists would need to know how effective such systems are and what system limitations may be. According to a 2004 contract, described on the Port of Oakland Web site, Pacific Northwest National Laboratory is designing and installing the RPMs for LA/LB.[17] Nevertheless, the machines' efficacy is debatable. The American Association for the Advancement of Science has published a skeptical report, and an expert of theirs used it to conclude, "More needs to be done to protect the United States from smuggled nuclear weapons because current portal monitors probably could not detect even a few kilograms of highly enriched uranium, even if only lightly shielded" (see Strengthening the Global Partnership, 2005).

---

[15] Govpro.com (2005); Ahern (2005); Interdict/RADACAD (undated). Examples of commercially available RIIDs include ExploraniumTM GR-135, GR-135CN made by SAIC, and the Thermo identiFINDER (WorldSecurity-index. com, undated[a], undated[b]).

[16] RPMs are detection devices that provide CBP officers with a passive, nonintrusive means to screen containers, vessels, or vehicles for the presence of nuclear and radiological materials. These systems do not emit radiation but are capable of detecting various types of radiation emanating from nuclear devices, dirty bombs, special nuclear materials, natural sources, and isotopes used in medicine and industry.

[17] Port of Oakland (2004, undated); WorldSecurity-index.com (2001). An example of the types of capabilities and technical details for one manufacturer can be found in Orphan et al. (2004).

MU-Vision, which manufactures radiation detection technologies, published a white paper (MU-Vision, undated) detailing some weaknesses of current x-ray and gamma-ray scanning technologies including passive gamma-ray and neutron detectors (e.g., TECO Electronics Co.'s "Radiation Pagers"), mobile and portal systems (from American Science and Engineering, Inc. [AS&E] or SAIC), two-dimensional transmission x-ray systems (e.g., AS&E, L-3 Security and Detection Systems, Inc., RapiScan), two-dimensional—transmission gamma-ray systems (e.g., SAIC Vehicle and Cargo Inspection System [VACIS]), and pulsed fast neutron analysis (PFNA) offered by Ancore Corporation. Analysis of such limitations provides at least some clues on hiding radiological materials from detection.

In a mathematical study, professors at Stanford University concluded that, based on existing security policies, protocols, and detection technologies, a nuclear warhead containing "4 kg of weapons-grade plutonium or 12 kg of weapons-grade uranium, shielded with tungsten and lithium hydride, and which is shipped within a inter-modal container with other cargo" (Flynn, 2004) has only a 9.75-percent chance of being detected if shipped in a container from a trusted shipper.[18] Taking into account the added security and protocols associated with being a certified shipper, the study estimates that the probability of detecting a smuggled nuclear warhead increases to a maximum of 24 percent.[19] Added Stephen Flynn (2004), an expert on maritime shipping and port security,

> If the weapon is placed in a 20-foot container which is commonly used to move heavy machinery, the probability of detection drops to nearly zero because the radiography cannot penetrate cargo that would likely be between the wall of the container and the weapon.

Current events also cast doubt on the system's ability to detect illicit cargo. In addition, in January 2005, 32 Chinese stowaways were found in two cargo containers in the Port of Los Angeles after surviving a two-week ocean journey from Hong Kong. No form of x-ray or other scanning device detected the stowaways. They were identified, according to Agence France Presse (2005), when a crane operator spotted three men climbing out of the cargo container.

Security Forces

The CBP bureau receives an electronic bill of lading, or manifest data, for approximately 98 percent of the sea containers at least 24 hours before they arrive at U.S. seaports. These data are combined with other intelligence data and processed via CBP's automated targeting system to identify low-risk and high-risk cargo (CBP, 2005).

Cargo is designated as low-risk if it is being shipped by established and trusted importers who are registered with CBP; by one estimate, such cargo is five to eight times less likely to be stopped for security screening or inspections (Samuel Shapiro and Company, undated). Shippers may participate in the DHS C-TPAT program by assessing themselves using secu-

---

[18] Flynn (2004); Lane (2004). The article also quotes Steven Fetter of the University of Maryland and a specialist on nuclear proliferation as saying that he was skeptical of the Stanford findings; however, he did suggest that challenges of detecting highly enriched uranium are "well taken."

[19] The term *certified shipper*, as used in the original article by Wein et al. (2006), refers to a shipper that has undergone a self-certification as dictated by the C-TPAT program and gains the status of a certified shipper.

rity guidelines established by DHS and trade associations and agreeing to implement certain security reforms (CBP, undated). CBP is obligated to audit and verify the C-TPAT member. An article published in *USA Today* on May 25, 2005, stated that, according to the GAO and the U.S. Senate Permanent Subcommittee on Investigations, customs officials have verified as secure only 11 percent of the 4,357 importers certified as C-TPAT members.[20] In the same article, a DHS representative claims that the actual number of shippers that have been verified as secure is 30 percent.[21]

Terrorists may know who the trusted shippers (C-TPAT members) are (Flynn, 2002), despite the fact that CBP does not circulate this list.[22] Although it was not possible to find a complete list, brief searches on the Internet did reveal information about some firms that are participating in the C-TPAT program, such as Wal-Mart, Target, and Maersk.[23]

The CBP CSI has been established with foreign ports that send substantial container traffic to the United States. CSI's purpose is to inspect high-risk cargo—before it begins its voyage to U.S. ports—and to enhance information sharing. It does so by deploying gamma-ray or x-ray and radiation detection equipment and personnel to foreign ports of embarkation (Embassy of the United States, 2004). Thirty-three ports, accounting for two-thirds of all container traffic destined for the United States, were registered as CSI ports as of May 2005; they include 19 of the top 20 foreign ports by tonnage.

Should terrorists be concerned about C-TPAT and CSI? According to Senate testimony from GAO,

> [approximately] 35 percent of U.S.-bound shipments from CSI ports were not targeted and not subject to inspection overseas—the key goal of the CSI program. In addition, as of September 11, 2004, 28 percent of the containers referred to host governments for inspection were not inspected overseas for various reasons such as operational limitations. (Stana, 2005)

According to a *USA Today* article, only about 17.5 percent of containers deemed high risk were inspected at the port of departure (Hall, 2005).

The ports listed in Table A.2 include the 20 foreign ports that ship the largest volume of ocean containers to the United States and accounted for approximately 66 percent of all con-

---

[20] Hall (2005). The article refers to information contained in Stana (2005b).

[21] According to an interview that was published after this project's information collection phase, the director of the C-TPAT program states that customs officials have verified the security procedures of 1,405 of 5,636 (approximately 25 percent) C-TPAT members. See Tirschwell (2006).

[22] From a FAQ on the CBP Web site:

QUESTION: Why can't CBP distribute a list of the C-TPAT partners?

ANSWER: C-TPAT is a unique partnership between CBP and the C-TPAT trade partner signatory. It is a voluntary program that is not regulated by law. For security and confidentiality purposes, CBP will not share any information regarding C-TPAT application or partnership status with anyone outside the company's authorized officials. (C-TPAT, undated)

[23] Although a complete list of C-TPAT members was not located, searches of individual shippers did provide information on C-TPAT participants, including Wal-Mart, Target, Maersk, and MillenniaPartners (Judd, 2006; Roberti, 2004; Maersk Logistics, undated; MillenniaPartners, 2004).

tainers that arrived in U.S. seaports in 2001.[24] The list also includes ports that may ship less to the United States but may have terrorism related to geographic concerns (Stana, 2005b).

Table A.2
GAO Listing of CSI Operational Seaports (as of February 2005)

Country/Region	CSI Port	Date CSI Operations Began at Port
Canada	Halifax	March 2002
	Montreal	March 2002
	Vancouver	February 2002
The Netherlands	Rotterdam	September 2002
France	Le Havre	December 2002
	Marseilles	January 2005
Germany	Bremerhaven	February 2003
	Hamburg	February 2003
Belgium	Antwerp	February 2003
	Zeebrugge	October 2004
Republic of Singapore	Singapore	March 2003
Japan	Yokohama	March 2003
	Tokyo	May 2004
	Nagoya	August 2004
	Kobe	August 2004
Hong Kong Special Administrative Region of China	Hong Kong	May 2003
United Kingdom	Felixstowe	May 2003
	Liverpool	October 2004
	Southampton	October 2004
	Thamesport	October 2004
	Tilbury	October 2004
Italy	Genoa	June 2003
	La Spezia	June 2003
	Livorno	December 2004

---

[24] Most cargo is transported through a relatively small number of ports, such as those listed in Table A.2. However, according to *Lloyd's List, Ports of the World 2004*, approximately 756 ports worldwide that the capacity to handle general cargo and container traffic.

Table A.2—Continued

Country/Region	CSI Port	Date CSI Operations Began at Port
	Naples	September 2004
	Gioia Tauro	October 2004
South Korea	Busan	August 2003
South Africa	Durban	December 2003
Malaysia	Port Klang	March 2004
	Tanjung Pelepas	August 2004
Greece	Piraeus	July 2004
Spain	Algeciras	July 2004
Thailand	Laem Chabang	August 2004

SOURCE: Stana (2005b).

The Department of Energy (DOE) administers a separate program, the Megaports Initiative, that seeks to reduce the risk of radioactive materials being used against the United States or its allies by providing radiation detection to foreign government personnel at key international seaports to screen shipping containers entering and leaving those ports (National Nuclear Security Administration, undated). However, according to the DOE Web site, the Megaports Initiative is only operational in Greece, Bahamas, Sri Lanka, and the Netherlands and is at various stages of implementation in 10 other countries. In addition, a 2005 GAO report mentions operational and technical challenges related to implementing the detection equipment and environmental factors, such as sea spray and high winds, that affect detector performance or long-term sustainability (U.S. Government Accountability Office, 2005a). In addition, once the equipment is installed and control is handed over to the host government, it becomes more difficult to verify the effectiveness of the equipment and the procedures for their operation.

Finally, terrorists who watch television may already understand how easy it is to ship radioactive material into the United States. An ABC news team successfully found a shipper to send depleted uranium from Indonesia to LA/LB undetected. "We did not tell the company about the depleted uranium," said one of the producers, "and they never asked" (Ross, 2004). Maersk Logistics (Denmark) handled the shipment. Their procedures did not require their agents to inspect containers loaded outside of the pier area at the Port of Jakarta, Indonesia; their "door-to-door service" allowed the container to be loaded at a furniture store there in Jakarta. Maersk security official, John Hyde, responded, "We rely on screening of government authorities to validate shipping contents" (Ross, Schwartz, and Scott, 2002). Following the ABC News article, Maersk vowed to review its security procedures.

What the Expert Searcher Found

Many relevant standards, including the IEEE N42.35 (American National Standard for Evaluation and Performance of Radiation Detection Portal Monitors for Use in Homeland Security) (American National Standards Institute and Institute of Electrical and Electronics Engineers, 2004) and the draft American National Standards Institute (ANSI) 42.38, establishes standards for spectroscopic radiation portal monitors.[25] Given the radiation level (technically referred to as the "source activity") listed and with some knowledge of nuclear materials, it is possible to determine the minimum quantity of material the RPMs are being established to detect successfully. Therefore, in theory, with that and other available information or knowledge, it would be possible to design a container for shipping a nuclear device that would fall below the published minimum standards for detection.

A DHS bulletin described the findings from four reports looking at the effectiveness of commercially available nuclear detectors (see Mayer, 2005). According to the summary published by DHS, "none of the radiological and nuclear detection equipment passed all of the tests" (Mayer, 2005). This suggests that current commercially available devices have significant limitations in detecting various classes of radiological devices.

A 2004 DoD report that discusses the performance of current radiation detectors states that plutonium devices can be detected in vehicle portals, cargo containers, and moving vehicles *if the device is unshielded or lightly shielded* (U.S. Department of Defense, 2004). The report also states that the detection of highly enriched uranium (HEU) is very difficult and is limited by the short range over which radiation detectors operate. It also states that lightly shielded devices can be detected at radiation detection portals (through which vehicles or containers pass) but, in other cases, the HEU devices can only be detected if they are unshielded.

Summary Findings for Scenario 3

Public information indicates that no comprehensive mechanism for scanning incoming cargo at LA/LB is likely to be in place until the end of 2005, if then. Even at that point, the current technology would be limited by its sensitivity and how it is employed (on moving trucks). Another risk factor that terrorists may face is not knowing whether the details of a cargo manifest would subject the cargo to additional scrutiny. Enough anecdotes could be gathered on that issue to make an educated guess as to what criteria are used. Public information suggests that many foreign ports are unlikely to do much inspection on their own. U.S.-driven CSI initiatives will improve detection capabilities at some foreign ports (identified above). Yet, in addition to having to contend with the same technical limitations present in scanning equipment used in U.S. ports (according to an April 2005 GAO report), the CSI program is not sufficiently staffed to inspect all U.S.-bound cargo and must deal with additional diplomatic constraints (e.g., host government's permission) (Stana, 2005a).

---

[25] Spectroscopic RPMs are able to distinguish different radiation sources (Homeland Security Advanced Research Projects Agency, undated).

Scenario 4:
Madrid-Style Bomb Attack on Commuter Train in the NYC East River Tunnel

In this scenario, terrorists board the LIRR at Douglaston Station at 8:10 a.m. on a normal weekday—a time of high commuter passenger traffic.[26] They carry explosive devices. Each takes a seat on a different car. They intend to depart the train at Woodside Station at 8:26 a.m., leaving behind their explosives, which are timed to go off at 8:31 a.m. (seven minutes before the train is due at Penn Central Station). If the trains are running on time, the explosions would take place when the train is inside the East River Tunnel.

Critical Information for the Mission When No Other Information Has Yet Been Collected

- Security measures: These include security monitoring devices and sensors that might detect the presence of explosive materials.
- Security forces: These include security presence and procedures at the train station and on the train itself. Detection by security forces could negate the operation or minimize its consequences.
- Target: Execution of the operation requires a minimum amount of information about avenues of approach to the target. Maximizing an attack's consequences requires information on train schedules, on-time averages, passenger density information, and some structural details of the East River Tunnel.

Our findings are presented in three categories.

Security Measures

Officials at LIRR stated that "improved electronic access control and increased security guard services" were being added to their facilities and that surveillance cameras and intrusion alarms were being upgraded. However, as of March 7, 2005, the railroad had not met the March 2004 DHS directives to install bomb-resistant trash cans, to erect vehicle and pedestrian barricades, and to install closed-circuit security cameras.[27] Publicly available images of the Douglaston Station fail to indicate any security or monitoring equipment (Long Island Rail Road, undated; Long Island Rail Road History, undated). Figure A.2 is a map of the LIRR.

---

[26] The LIRR was chosen because it is both the oldest and most heavily used commuter railroad in the country. It has also been criticized recently for lack of security (Schumer, 2004a).

[27] Schumer (2005). See also a news release regarding the directive at U.S. Department of Homeland Security (2004).

Figure A.2
Station Map of the MTA Long Island Rail Road

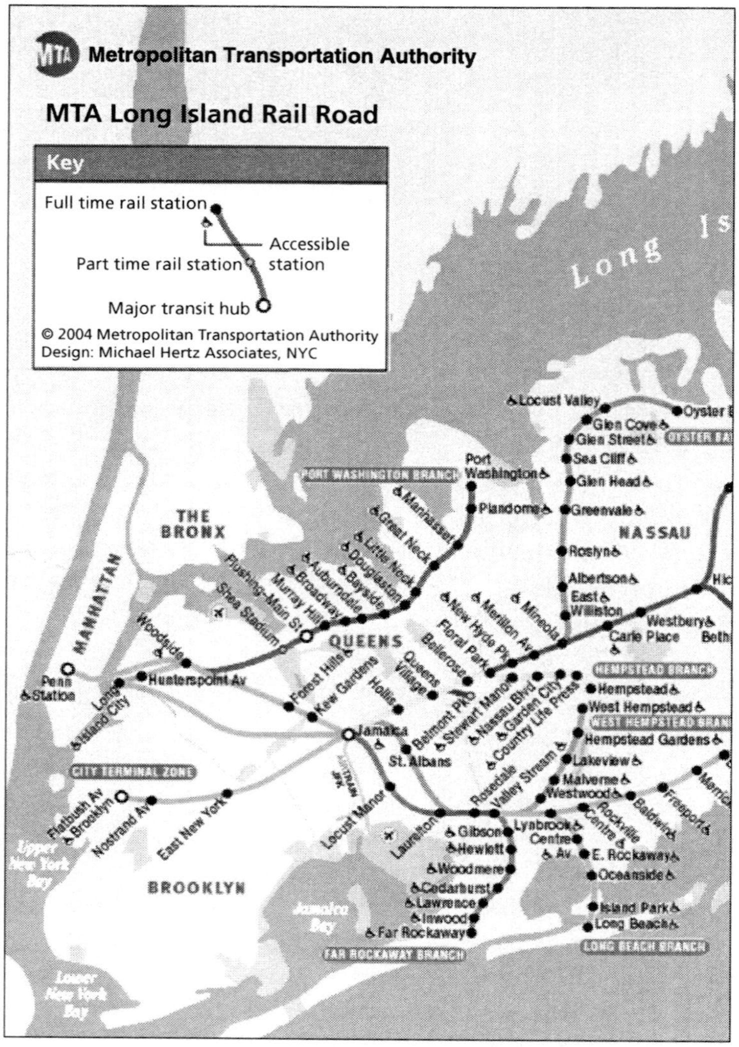

©Metropolitan Transportation Authority. Used with permission.
SOURCE: Metropolitan Transit Authority (2004).
RAND TR360-A.2

Shortly after the London bombings in July 2005, the Port Authority of New York and New Jersey disabled cell phone communications[28] were disabled for approximately two weeks in the four tunnels leading into Manhattan from the west (Wallace et al., 2005). Such countermeasures are ineffective against other methods (e.g., suicide bombings) and mechanisms, including mechanical timers, and other radio-based signaling (United Nations Office on Drugs

---

[28] Verizon digital and analog service provides cell phone coverage is provided on the LIRR under the East River. See WirelessAdvisor.com Forums (undated[a], undated[b]).

and Crime, undated). A July 2005 *New York Sun* article noted recent requests for proposals from MTA for installation of systems that would enable cell phone reception while preventing cell phones from being used as triggering devices (Smerd, 2005).

Security Forces

As noted, the MTA provides security for the LIRR; 692 MTA officers "patrol Metro-North and LIRR stations and the bridges and tunnels" (Donohue and Gittrich, 2004). An article in the *National Corridors Initiative, Inc.* online magazine quoted James Dermody, president of the LIRR, as saying that the MTA police were placing "special emphasis" on critical locations and possible "high-value targets" along the MTA (King, 2004). Dermody describes these high-value targets as locations "where [there] is the most potential [for] loss of life, serious economic impacts to the region, high . . . recovery or replacement [costs], or large [. . .] environmental damage." In the article, Dermody mentions examples "such as Penn Station, Jamaica [Station], the East River and the Atlantic[29] tunnels" (King, 2004) as locations that MTA police considered to be critical or high-value targets.

Several Web sites provide information on the police frequencies used by the LIRR (Long Island Area Scanning Resources, 2001b); Table A.3 lists some of them.

Target

As mentioned above, the images available from the MTA Web site as well as other locations provide useful information about Douglaston Station and the absence of any security

Table A.3
MTA Police Frequencies

Frequency	Agency	Description
160.455	MTA	Police F1
160.605	MTA	Police F2
160.320	MTA	Police F3
452.6375	MTA	Police Portables
452.6875	MTA	Police Portables
452.8125	MTA	Police Portables

SOURCE: Long Island Area Scanning Resources (2001).

---

[29] The reference in this quote is likely to the Atlantic Avenue Tunnel, also known as the Cobble Hill Tunnel. The tunnel was first opened in 1844 and ran for approximately 2750 ft. underneath Brooklyn, New York, from Hicks Street to Boerum Place. The tunnel was closed and both ends of the tunnel sealed shut in 1861. Since then, the tunnel has been reopened and inspected on a couple of occasions because of concerns that the tunnel was being used for staging criminal or terrorist activities. For example, in 1916 the Federal Bureau of Investigation inspected the tunnel because of concerns that German terrorists were using the tunnel to build bombs, then again because of concerns over the tunnel being used to house bootleg whisky stills in 1920s. More information can be found at NationMaster.com (2005).

monitoring or barriers to approaching the train. The open-platform format suggests easy access. Figure A.3 shows the Douglaston Station platform.

The Washington Port branch of the LIRR, the line on which the Douglaston Station is located, is serviced by either M1 or M3 electric cars manufactured by the Budd Company, General Electric or Bombardier. Some information about the interior of types of cars was located, including images of the interiors. The interiors of both the M1 and M3 electric cars[30] are very similar. The M1 and M3 cars are being replaced by the M7 electric cars, which also have various photos of the interior available publicly, again showing layout and possible locations to place an explosive device (railfanpete, 2004).

Train schedules for the LIRR are available on the World Wide Web (Long Island Rail Road, 2005a). On-time rates are also reported, but since "on time" is defined as no more than 5 minutes and 59 seconds late, the published rates are not of sufficient detail to time detonations under the East River Bridge (Long Island Rail Road, 2005b). Beyond vague indications of crowding during rush hour, the red team was unable to discover any operationally useful information about variations in passenger density. The red team also failed to find any structural details about the East River Tunnel.

Figure A.3
Photograph of Douglaston Station

©Metropolitan Transportation Authority. Used with permission.
SOURCE: Long Island Rail Road (undated).
RAND *TR360-A.3*

---

[30] Examples of images found of the interiors of the M1, M3, and other LIRR trains can be found at the Web site IND/BMT/IRT (undated).

Summary Findings for Scenario 4

The red team did not find any publicly available data suggesting there are any countermeasures in place to thwart the attack scenario outlined above. Data do suggest that Douglaston Station is easily accessible and that the main efforts of security forces are focused elsewhere. Key data relevant to maximizing the consequences of an attack of this kind were unavailable, as was detailed information regarding the likelihood of encountering security forces. Similarly, the red team was unable to uncover any information that would allow it to predict with confidence whether the "abandoned parcels" containing the bombs would be allowed to remain undisturbed long enough to reach their target and detonate.

Scenario 5: MANPADS Attack on a Flight Bound into LAX

This scenario involves a terrorist who fires a MANPADS at a passenger airliner landing at LAX.[31]

Critical Information for the Mission When No Other Information Has Yet Been Collected

- Security measures: These might be any system on the ground or on the plane that might prevent the missile from hitting the plane or any systems for observing possible launch points.
- Security forces: These include indication of security force patrols or activities that might lead to the discovery of the attacker before the attack is launched.
- Target: This is the predictable path of the target (flight path), information for identifying a suitable location to shoot at the intended target, and information necessary to arrive there.

Security Measures

A 2004 trade article stated that commercial airlines had no defense against MANPADS-type attacks.[32] Despite a recent RAND report (Chow et al., 2005) discussing the prohibitive economics of defending aircraft against MANPADS, starting in August 2005, DHS will begin testing antimissile equipment manufactured by Northrop Grumman and BAE Systems on three airliners. According to the article, the "system fits inside a pod that bolts to the bottom of a jet and is equipped with sensors that can detect a shoulder-fired missile. A swiveling turret would then fire a laser beam that could confound the sensitive heat-seeking components of the missile" (Stoller, 2005).

---

[31] Although the scenario specified an inbound flight in order to guide the red team's information search, the impact of hitting an outbound flight is not greatly different. Terrorists contemplating one or the other need similar but not identical information.

[32] Allied Pilots Association (2004). For a technical compendium of potential defenses, see material contained in DoD's budget justification for its FY 2004 research, development, test, and evaluation budget activities (U.S. Department of Defense, 2003).

Limited information was found on countermeasures that might detect attackers prior to MANPADS launch. The red team did find a December 2004 report that indicated an expanded effort to protect LAX from missile threats ("Security Boosted at LAX to Guard Against Missiles," 2004). Listed countermeasures included "expanded helicopter surveillance, new perimeter fencing, stepped-up police patrols and additional training to help authorities identify such weapons" ("Security Boosted at LAX to Guard Against Missiles," 2004).

Security Forces

The red team was able to find some details of the LAX police department's force size and capabilities. For example, a March 2005 *San Diego Union Tribune* article notes that

> the LAX Police Department has an authorized staff of 354 sworn officers and 318 civilian traffic and security officers, and the airport also pays for 37 to 44 LAPD officers who are permanently stationed at the department's LAX substation near Terminal 8. Additionally, records show that the airport pays all or part of the costs of about 65 other officers and detectives from special units, including canine, narcotics, bomb squad, traffic, forgery and anti-terrorism. (Gregor, 2005)

However, little was found regarding security efforts outside the perimeter of the airport proper beyond what was noted above under countermeasures.

A publicly available 2004 RAND briefing recommends increased deployment of plain-clothes security in autos, boats, and so forth, to patrol areas where MANPADS might be launched, with particular emphasis on vantage points where terrorists might be in range to strike planes "mid-take off and landing" (Stevens et al., 2004). The fact that this recommendation was made led the red team to conclude that, at least prior to the recommendation, security forces were either not deployed at all or were not adequately deployed in this fashion. The red team sought and failed to find any indication that action had been taken on this recommendation.

Target

A great deal of information on incoming flight paths to LAX is publicly available. This includes near–real-time information (delayed approximately 10 minutes) about incoming and outbound flights from LAX (iflylax.com, undated) as well as instrument approaches and runways for every major airport (AirNav.com, undated). This would allow terrorists to target a specific flight (if they were so inclined) or observe regular flight paths and schedules over possible launch areas without actually being near the launch area.

Identifying an ideal launch point based on publicly available data proved to be surprisingly difficult. Although ubiquitous map sites make street maps and detailed overhead imagery of candidate launch areas easily available, the red team was unable to find sufficient information to evaluate the quality of lines of sight from candidate launch points to the flight paths of arriving planes or to assess the extent to which these candidate launch points are exposed to view by facility or local security forces.

Summary Findings for Scenario 5

Publicly available information indicates that planes landing at LAX are not presently equipped with any countermeasures for a MANPADS attack. Information indicating efforts to police the area surrounding LAX for such attacks was uncovered, but the red team was unable to estimate the effectiveness of these efforts. Selecting a launch point that had both good lines of sight to target flight paths and reasonable cover from casual observation and observation by security forces proved difficult to find through publicly available off-site sources. However, it does not seem to be too far-fetched to imagine such sites being easily selected with on-site surveillance that is actually more "near-site" than on-site, given that such reconnaissance would be well away from the perimeter of LAX proper.

Scenario 6:
Suicide Boat Rams a Docked Cruise Ship at the Port of Los Angeles

In this scenario, a terrorist cell loads a yacht, moored at the Cabrillo Yacht Marina (Port of Los Angeles) (California Yacht Marina, undated), with four tons of explosives. One morning, as a cruise ship nears its docking port, this private yacht proceeds at full throttle through the port's main channel and hits the side of the cruise ship. The explosives detonate, leaving a gaping hole at the waterline amidships.[33]

Critical Information for the Mission When No Other Information Has Yet Been Collected

- Security measures: These are countermeasures to prevent or mitigate the attack.
- Security forces: These are security procedures around the target. What are the chances that security forces can detect and then disable or intercept the bomb boat before it reaches the target?
- Target: These include the location of and paths to the target. To maximize consequences, structural details of the target are required.

Security Measures

The City of Los Angeles (undated) Hazard Mitigation Plan mentions a project involving the development of waterborne perimeter security barriers. To prevent boats from entering certain areas, such barriers would be placed at the LA World Cruise Center, the Long Beach Cruise Terminal, the Catalina Express ferry terminals in the ports of both LA and Long Beach, the General American Transportation (GATX) liquid petroleum gas (LPG) terminal in LA and the Arco terminal in Long Beach, among other sensitive terminals. Designs for these counter-

---

[33] Although the scenario specified an attack on a cruise ship, other similar scenarios, such as attacks on smaller commercial vessels, may result in more fatalities or greater economic consequences. For example, in February 2004, the southern Philippines–based Abu Sayyaf detonated an explosive on a ferry vessel, killing 100 people. Terrorists contemplating one or the other need similar but not identical information is needed by terrorists contemplating one or the other (Luft and Korin, 2004).

measures were part of a grant awarded from DHS in November 2004 (Port of Los Angeles, 2004). No information was found suggesting that these countermeasures were actually in place by July 2005.

Security Forces

According to one article (a FAQ on cruise lines), "[e]very U.S. port now maintains and enforces a minimum 300-foot 'no float zone,' a security perimeter that prohibits private craft from coming near cruise ships.[34] In addition, cruise ships are getting an armed U.S. Coast Guard (USCG) escort in and out of port" (Rubacky, undated). This is confirmed by an April 2003 blogger who noted a USCG escort and its defensive behavior toward an errant fishing vessel as her cruise ship departed San Francisco (Adamec, 2003). A separate article mentions other security procedures that have been used to prevent possible terrorist attacks, including dive teams and dog teams to check boxes for explosives (Walsh, 2002). It is unclear exactly how many USCG vessels accompany cruise or other ships as they enter, dock, and leave ports. According to one article, only one USCG vessel is mentioned as escorting a cruise ship into the Port of Miami (Kitfield, 2002). Information on USCG assets at LA/LB is available from the official USCG Web site (U.S. Coast Guard Sector Los Angeles–Long Beach, undated).

The red team also identified a USCG Web site that listed the assets located at the USCG small boat station at LA/LB. Its assets include three utility boats, two 24-ft. nonstandard boats, one 25-ft. inflatable collar SAFE boat (Safeboats International, undated), one 23 ft. inflatable collar SAFE boat, and one Homeland Security Response Boat (RB-HS) (Kennedy, 2003). A different article also notes, "The new 25-foot response boats will replace nearly 300 nonstandard shore-based craft. They are more maneuverable than the older boats. Outfitted with twin engines, they are capable of speeds in excess of 40 knots" (Kennedy, 2003).

USCG forces also use helicopters armed with sharpshooters for intercepting drug smugglers; it is unclear whether this capability is available for counterterror as well.

> If the suspect vessel fails to stop after these numerous visual and verbal warnings, the helicopter crew will take up a firing position alongside the go-fast and fire warning shots across their bow to further compel them to stop. If the warning shots do not convince the suspects to stop, the helicopter crew prepares to disable the vessel by shooting out the go-fast's engines. Using precision, laser-sighted .50 caliber rifles, the helicopter crew positions themselves alongside the fleeing go-fast for disabling shots.[35]

Target

Full overhead imagery is available for LA/LB, including images showing a cruise ship parked at a dock. Overhead imagery provides no evidence of barriers afloat around the depicted cruise

---

[34] This statement is consistent with current regulations in U.S. Code Title 33, Section 165, Regulated Navigation Areas and Limited Access Areas, that describe the designation of security and safety zones in and around specific U.S. ports (U.S. Code, 2003).

[35] Neubecker (2003). However, a boat intent on detonating alongside a cruise ship could travel the 100-yard no-float zone in a matter of seconds. Therefore, unless the helicopter is already in position, it is unlikely to be an effective countermeasure.

ship. On the Port of Los Angeles Web site are links to the various cruise lines use as their beginning and end destination (Port of Los Angeles, undated). From these links, one can easily find out the dates and times of departure and arrival. For example, Royal Caribbean cruise ships usually begin boarding on day of departure around 5:00 p.m. and end trips with an arrival time in the morning around 7:00–8:00 a.m. The red team did not find useful structural information about cruise ships, such as the hull thickness, the location of bulkheads, or other relevant engineering features.

Summary Findings for Scenario 6

The red team discovered considerable information regarding USCG assets at LA/LB and some details about their performance. Publicly available off-site sources suggest that countermeasures such as floating barriers do not currently exist. Information regarding deployments of USCG assets and information sufficient to assess the USCG forces' ability to intercept and prevent the attack was lacking.

Crosswalk of ModIPB and al Qaeda Manual

This appendix provides a side-by-side comparison of the information categories described in the al Qaeda manual, corresponding to Tables 2.1 through 2.3 and the categories identified within the ModIPB as described in Tables 2.4 through 2.7. The doctrinal U.S. Army IPB process identifies four steps for intelligence collection and analysis. They are (1) define the battlefield environment and identify the boundary of the operational area; (2) describe the battlefield effects and determine how the environment will affect threat and friendly operations; (3) evaluate the threat and determine the capabilities, doctrine, tactics, techniques, and procedures that threat forces may employ; and (4) determine the threat COAs and integrate the information from the previous steps to create meaningful COAs. The IPB doctrine describes the types of activities that should occur during each step and provides intelligence collectors and analysts with a framework for identifying intelligence requirements for specific missions or objectives. However, it does not provide a discrete list of information categories to be collected that can be succinctly compared with the al Qaeda manual.

Table B.1
Comparison of the ModIPB Information Categories and the Information-Gathering Requirements Identified by the al Qaeda Manual

ModIPB	al Qaeda Manual: Exterior Information-Gathering Requirement (Table 2.1)	al Qaeda Manual: Interior Information-Gathering Requirement (Table 2.2)	al Qaeda Manual: Information Requirement About Bases or Camps (Table 2.3)
1. Avenue of approach and ease of access			
Location of target			Location
Surrounding terrain and buildings	Characteristics of the area around the place		Space [area]; exterior shape
Maps	The area, physical layout, and setting of the place		
Blueprints		Electric box, inside parking, telephone lines and switchboard location, number of floors and rooms	Telephone lines and means of communication

75

Table B.1—Continued

ModIPB	al Qaeda Manual: Exterior Information-Gathering Requirement (Table 2.1)	al Qaeda Manual: Interior Information-Gathering Requirement (Table 2.2)	al Qaeda Manual: Information Requirement About Bases or Camps (Table 2.3)
Types of building construction			
Critical points	Nearby embassies and consulates		
OCOKA			
Available paths to target	Transportation means to the place		Transportation to it
Exact path(s) to take			
Go/no-go areas (because of barriers, obstructions, or impassable terrain)	How wide are the streets and in which direction do they run leading to the place?		
Areas of restricted or limited access (security restrictions)			
Rules and laws governing movement (vehicular or otherwise) in target area	Traffic signals and pedestrian areas		Leave policy
Traffic conditions (all relevant vehicular and pedestrian modes)	Traffic congestion times		
2. Target			
Possiblie locations from which to launch the attack			
Mobility and variability of the target; if mobile, (predictable) paths it takes			
Relevant features and structure of the target (i.e., technical details)			
3a. Security forces			
Locations of headquarters, stations, checkpoints		Number and location of guard posts	Guard posts, fortification, and tunnels
Overall size and types (uniformed, plainclothes, canine)			Brigades and names of companies
Number on duty at any one time, hours of duty, variation by times of day		Number and names of the leaders	Number of soldiers and officers; commander's name, rank, and arrival and departure times; sleeping and waking times (presumably of troops or security forces)

Table B.1—Continued

ModIPB	al Qaeda Manual: Exterior Information-Gathering Requirement (Table 2.1)	al Qaeda Manual: Interior Information-Gathering Requirement (Table 2.2)	al Qaeda Manual: Information Requirement About Bases or Camps (Table 2.3)
Applicable operational jurisdictions, number of security forces contributed by each			Unit using the camp
Capabilities			Weapons used
On what radio frequencies are they operating?			Ammunition depot locations
ROE or use-of-force policy			
Specific or individual deployments	Security personnel centers and nearby government agencies		
Fixed positions			
Patrols (routes, schedules, number of personnel, vehicles patrolling)			
Times of observations (cameras, live operatives)			
Number of security personnel required to be passed			
Variations by times of day			
Security plans (operational details)			Degree and speed of mobilization
Security response plan			
Past performance with previous (similar?) incidents			
Behaviors, plans, and capabilities at different alert levels			
Response times			
3b. Security measures			
Kinds of checkpoints to be passed			
Search procedures (For what will officials be looking or asking?)			
Cameras, scanners, and detection equipment operating over the area to be traversed			

Table B.1—Continued

ModIPB	al Qaeda Manual: Exterior Information-Gathering Requirement (Table 2.1)	al Qaeda Manual: Interior Information-Gathering Requirement (Table 2.2)	al Qaeda Manual: Information Requirement About Bases or Camps (Table 2.3)
Sensitivity of these devices			
Frequency with which sensors are read			
Illumination	Amount and location of lighting		Amount and periods of lighting
Specific countermeasures (e.g., vehicle barriers)			
3c. Other population groups			
Other people at the facility (Why are they there? What are they doing?)	Economic characteristics of the area	Number of people inside	
Bystanders, recreational users, passengers			
Differences in population at different times of day			
Vigilance instructions and emergency phones			
Level of security training for nonsecurity employees at the facility			
Schedules of regular arrivals and departures from target area		Individuals' times of entrances and exits	
Ease of camouflaging oneself as a member of one of the population groups			
4. Threats to the operation			
Threat posed by security forces and measures			
Deployments, response times, vehicles, equipment, plans			
Cascading information (from organization of oversight and headquarters locations to who will be on the avenue of approach on attack day)			

Table B.1—Continued

ModIPB	al Qaeda Manual: Exterior Information-Gathering Requirement (Table 2.1)	al Qaeda Manual: Interior Information-Gathering Requirement (Table 2.2)	al Qaeda Manual: Information Requirement About Bases or Camps (Table 2.3)
Estimated effectiveness of security response capabilities (including communications			
Threat posed by employees of nonsecurity facilities			
Citizens (concentrations, heightened vigilance)			
Weather (as it impacts effectiveness of the operation)			

Bibliography

"9/11 Recordings Chronicle Confusion, Delay," *CNN.com*, June 17, 2004. As of June 19, 2006:
http://cgi.cnn.com/2004/ALLPOLITICS/06/17/911.transcript/

"42nd Street/Times Square Mega-Complex," *Station Reporter*, undated Web page. As of August 1, 2005:
http://www.stationreporter.net/timessq.htm

Adamec, Justene, "The Coast Guard," *Calblog*, April 18, 2003. As of August 3, 2005:
http://www.calblog.com/archives/001790.html

Agence France Presse, "32 Chinese Stowaways Found in Container in Los Angeles," January 17, 2005.

Ahern, Jayson P., "Statement by Jayson P. Ahern, Assistant Commissioner, U.S. Customs and Border Protection, Hearing on 'The Security of the Nation's Cargo as It Enters United States Ports and the Methods Used by Government and the Private Sector to Combat the Smuggling of Illegal and Potentially Dangerous Cargo into the United States," March 15, 2005. As of August 2, 2005:
http://judiciary.house.gov/media/pdfs/ahern031505.pdf

AirNav.com, "Airport Information," undated Web page. As of August 3, 2005:
http://www.airnav.com/airports/

Air Safety Week, "Report: Terrorist Bombs, Not Missiles, the Deadliest Threat," October 4, 2004. As of July 28, 2005:
http://www.findarticles.com/p/articles/mi_m0UBT/is_38_18/ai_n6283163/pg_2

Allied Pilots Association, *Flightline*, Summer 2004. As of August 3, 2005:
http://www.alliedpilots.org/Public/Publications/Flightline/2004_summerp.pdf

American National Standards Institute, and Institute of Electrical and Electronics Engineers, *American National Standard for Evaluation and Performance of Radiation Detection Portal Monitors for Use in Homeland Security*, New York: Institute of Electrical and Electronics Engineers, 2004.

Anderson, Ross, "Economics and Security Resource Page," undated Web page. As of November 2, 2005:
http://www.cl.cam.ac.uk/~rja14/econsec.html

Baker, John C., Beth E. Lachman, Dave R. Frelinger, Kevin M. O'Connell, Alexander C. Hou, Michael S. Tseng, David T. Orletsky, and Charles W. Yost, *Mapping the Risks: Assessing the Homeland Security Implications of Publicly Available Geospatial Information*, Santa Monica, Calif.: RAND Corporation, MG-142-NGA, 2004. As of November 21, 2006:
http://www.rand.org/pubs/monographs/MG142/

Baker, Stephen, "Looking for a Blog in a Haystack," *Business Week*, July 25, 2005, p. 38.

Bavak Security Group, undated homepage. As of June 28, 2007
http://www.bavakbeveiligingsgroep.nl

Berrick, Cathleen A., *Transportation Security: Systematic Planning Needed to Optimize Resources*, Washington, D.C.: U.S. Government Accountability Office, GAO-05-357T, February 15, 2005. As of November 24, 2006:
http://www.gao.gov/new.items/d05357t.pdf

Biden, Jr., Joseph R., "Security Amtrak's Future: Lessons from Madrid," Washington, D.C., March 10, 2005. As of July 29, 2005:
http://www.biden.senate.gov/newsroom/details.cfm?id=233289&&

Button, David, "Question for Delta Dash 'Known Shippers,'" *FaunaClassifieds Reptile and Ambiphian Business Forums*, February 7, 2005. As of August 2, 2005:
http://www.faunaclassifieds.com/forums/archive/index.php/t-63782.html

Buxbaum, Peter A., "Closing the Gap in Cruise Ship Security," *McGraw-Hill Homeland Security*, September 2004.

Calhoun, Lisa, "Slashing Rates for Known Shippers: Air Freight's New Rules for 2005," *ExhibitCity News*, Vol. 12, No. 9, September 2005. As of August 1, 2005:
http://www.exhibitcitynews.com/current/freight_2.php

California Yacht Marina, "Cabrillo Marina: Location," undated Web page. As of August 3, 2005:
http://www.cymcabrillo.com/location.html

CalTrade Report, "Security Slack at Ports of Los Angeles, Long Beach," September 6, 2004. As of October 16, 2006:
http://www.caltradereport.com/eWebPages/front-page-1094550384.html

———, "LA LB Ports to Have Nuke Scan Capability," June 7, 2005.

Carrillo, Silvio, "New Air Cargo Rules Proposed," *CNN.com*, May 17, 2005. As of August 1, 2005:
http://www.cnn.com/2005/TRAVEL/05/17/air.cargo.security/

Casanovas, Karen, Executive Director, The Alaska Air Carriers Association, "Air Cargo Security Requirements, Notice of Proposed Rulemaking," letter to Docket Management System, Anchorage, Alaska, January 10, 2005.

CBP—*see* U.S. Customs and Border Protection.

Chan, Sewell, "M.T.A. Slow to Spend Money on Transit Security," *The New York Times*, July 9, 2005, p. A1.

Cho, Aileen, "Engineers Are Digging Deep to Rebuild New York's Subways," *Engineering News-Record*, April 12, 2004. As of August 1, 2005:
http://www.enr.com/features/transportation/archives/040412.asp

Chow, James S., James Chiesa, Paul Dreyer, Mel Eisman, Theodore W. Karasik, Joel Kvitky, Sherrill Lingel, David Ochmanek, and Chad Shirley, *Protecting Commercial Aviation Against the Shoulder-Fired Missile Threat*, Santa Monica, Calif.: RAND Corporation, OP-106-RC, 2005. As of November 22, 2006:
http://www.rand.org/pubs/occasional_papers/OP106/

City of Los Angeles, "TE-HAR-06 Waterborne Perimeter Security Barriers," *City of Los Angeles Hazard Mitigation Plan*, undated, p. 4B-6. As of October 17, 2006:
http://www.lacity.org/epd/pdf_lhmp/Sec4B_Terrorism.pdf

City of New York Police Department, "Operation Atlas," undated Web page. As of November 22, 2006:
http://www.nyc.gov/html/nypd/html/atlas.html

Civil Aviation Authority, *Aircraft Hardening Research Programme Final Overview Report*, London, December 2001. As of August 2, 2005:
http://www.caa.co.uk/docs/33/CAPAP200109.PDF

Coalition of Airline Pilots Associations, "CAPA: Stowaway in a Box Reveals Gaping Hole in Air Cargo Security," press release, Washington, D.C., September 10, 2003. As of August 2, 2005:
http://www.alliedpilots.org/Public/Topics/Issues/is_11.pdf

Coalition of Journalists for Open Government, "House Funders Critical of TSA, Handling of Sensitive Security Information," undated Web page. As of August 1, 2005:
http://www.cjog.net/headline_house_funders_critical_of.html

Continental Airlines Cargo, undated(a) homepage. As of August 1, 2005:
http://www.cocargo.com/cocargo/default.asp

————, "Continental QUICKPAK®," undated(b) Web page. As of August 1, 2005:
http://www.cocargo.com/cocargo/CargoPages/quick.asp

Council on Foreign Relations, "Responding to Chemical Attacks," 2004.

C-TPAT—*see* Customs-Trade Partnership Against Terrorism.

Customs-Trade Partnership Against Terrorism, *Frequently Asked Questions (FAQ'S) for the C-TPAT Status Verification Interface (SVI)*, undated. As of August 3, 2005:
http://www.cbp.gov/linkhandler/cgov/import/commercial_enforcement/ctpat/svi/sviFAQ.ctt/sviFAQ.doc

Dateline D.C. Column, "The Threat to Rail," February 27, 2005. As of July 29, 2005:
http://www.pittsburghlive.com/x/tribune-review/opinion/columnists/datelinedc/s_307322.html

Delta Airlines, "Known Shipper Request," undated Web page. As of October 16, 2006:
http://www.delta.com/business_programs_services/delta_cargo/cargo_flight_availability/cargo_forms_
applications/known_shipper_request/index.jsp

Democracy Now!, "Pakistani Immigrant Being Deported for Taking Pictures of NY Reservoir Speaks from Jail," July 1, 2004. As of January 19, 2006:
http://www.democracynow.org/article.pl?sid=04/07/01/142212

DHS—*see* U.S. Department of Homeland Security.

Disastercenter.com, *The Al Qaeda Manual*, undated. As of July 29, 2005:
http://www.disastercenter.com/terror

Donnelly, Sally B., and Viveca Novak, "Ashcroft Takes Aim at Cargo Security," *Time Online Edition*, September 27, 2003. As of July 29, 2005:
http://www.time.com/time/nation/article/0,8599,490594,00.html

Donohue, Pete, and Greg Gittrich, "Hunting Terror: Police Dogs to Ride Rails at Penn Sta. as Republicans Convene," *Daily News: The Front Page*, April 29, 2004.

Dunlap, David W., "Crossroads of the Whirl," *The New York Times*, March 28, 2004, p. 14. As of August 1, 2005:
http://www.wirednewyork.com/forum/showthread.php?t=4649

Elias, Bartholomew, *Air Cargo Security*, Washington, D.C.: Congressional Research Service, Library of Congress, 2003. As of November 24, 2006:
http://www.fas.org/sgp/crs/RL32022.pdf

————, *Air Cargo Security*, Washington, D.C.: Congressional Research Service, Library of Congress, 2005. As of August 1, 2005:
http://burgess.house.gov/UploadedFiles/Air%20Cargo%20Security%20(January%2013,%202005)%20.pdf

Embassy of the United States, Brazil, "U.S. Wants to Prescreen Most Cargo Containers from Overseas," August 25, 2004. As of November 24, 2006:
http://brasilia.usembassy.gov/index.php?action=materia&id=2669&submenu=1&itemmenu=10

Epstein, Edward, "Feinstein Introduces Air Freight Bill: Jet Cargo Bays Called Security Risk," *San Francisco Chronicle*, January 16, 2003. As of July 28, 2003:
http://sfgate.com/cgi-bin/article.cgi?f=/c/a/2003/01/16/MN90477.DTL

FedEx Express, *Improvements in Air Cargo Dangerous Goods Safety*, briefing given at the NTSB Air Cargo Safety Forum at the NTSB Academy, Ashburn, Va., March 30–31, 2004. As of August 2, 2005:
http://www.ntsb.gov/events/symp_air_cargo/presentations/2.3_FedEx.pdf

Flynn, Stephen E., *Meeting the New Threats: Calibrating Public-Private Sector Strategies,* transcript from Meeting the Homeland Security Challenge: Maritime and Other Critical Dimensions, Session II, Cambridge, Mass., March 25–26, 2002.

———, "The Limitations of the Current Cargo Container Targeting,'" Washington, D.C., March 31, 2004. As of June 28, 2007:
http://www.cfr.org/publication/6907/limitations_of_the_current_cargo_container_targeting.html

Frank, Thomas, "Aviation Security Debate," *Newsday, Inc.*, September 8, 2003. As of July 28, 2005:
http://www.ncbfaa.org/whatsnew/relatedreleases/aviationsecurity.htm

Gibson, Allen R., "TSA Contracts Show Explosives Detection Still Needs Work at Airports," HomelandDefenseStocks.com, December 2004.

Google Maps. As of August 3, 2005:
http://maps.google.com

Govpro.com, "Busiest Seaports to Monitor for Radiation by Year's End," June 7, 2005.

Gregor, Ian, "LAX, LAPD Clash on Security Staffing Needs at the Airport," SignOnSanDiego.com, March 23, 2005. As of August 3, 2005:
http://www.signonsandiego.com/news/state/20050323-0630-cnslax.html

Grove, Lloyd, "Terrorism, Photography, and Civil Rights," August 19, 2004. As of July 29, 2005:
http://scaryny.com/archives/2004/08/terrorism_photo.php

Guart, Al, "Step into WMD 'Hell' with City's Chem Cops," *The New York Post*, July 4, 2004, p. 22.

Hall, Mimi, "WMD Security at Ports Under Fire," *USA Today*, May 25, 2005. As of October 15, 2006:
http://www.usatoday.com/news/washington/2005-05-25-wmd-ports_x.htm?csp=15

Harford Reptile Breeding Center, "Reptiles Shipping FAQ," undated Web page. As of August 2, 2005:
http://www.pythons.com/ship.html

Ho, Soyoung, "Plane Threat: Terrorists Have Never Shot Down an American Passenger Jet with Surface-to-Air Missiles. But Only a Matter of Time," *Washington Monthly*, April 2003. As of July 29, 2005:
http://www.findarticles.com/p/articles/mi_m1316/is_4_35/ai_99988623

Homeland Security Advanced Research Projects Agency, *American National Standard: Performance Criteria for Spectroscopy-Based Portal Monitors Used for Homeland Security*, undated. As of August 3, 2005:
http://www.hsarpabaa.com/Solicitations/ANSI-N42.38-WD-F1a.pdf

Howell, D. Stewart, "Facility Providing Realistic Chemical Training," *Army Link News*, May 13, 1998. As of July 29, 2005:
http://www.fas.org/spp/starwars/program/news98/a19980513chemical.html

iflylax.com, "LAX Air Traffic," undated Web page. As of November 24, 2006:
http://www.iflylax.com/lax_airtraffic.html

IND/BMT/IRT, undated homepage. As of October 17, 2006:
http://www.freewebs.com/halfliner/lirrmnrr.htm

Interdict/RADACAD, "Training, International and Domestic Border Security Training Equipment, Personnel Radiation Detector (PDR)," undated Web page. As of August 2, 2005:
http://www.pnl.gov/interdict/training/equipment.html

Internet Archive, undated Web page. As of November 21, 2006:
http://www.archive.org

I-Team, WISH-TV, "Interview with Darrin Kayser," May 25, 2004.

Japan-101 Information Resource, "Sarin Gas Attack on the Tokyo Subway," undated. As of July 29, 2005:
http://www.japan-101.com/culture/sarin_gas_attack_on_the_tokyo_su.htm

Judd, Jason, "Unchecked: How Wal-Mart Uses Its Might to Block Port Security," *AFL-CIO*, April 2006, pp. 1–13. As of November 24, 2006:
http://www.aflcio.org/corporatewatch/walmart/upload/walmart_unchecked_0406.pdf

Kennedy, Harold, "U.S. Coast Guard Ratchets Up Port Security," *National Defense*, June 2003. As of August 3, 2005:
http://www.nationaldefensemagazine.org/issues/2003/Jun/US_Coast_Guard.htm

King, Leo, "From DHS: McDonnell Gets Top Amtrak Security Post," *Rail Is Real*, Vol. 5, No. 20, May 17, 2004. As of August 3, 2005:
http://www.nationalcorridors.org/df/df05172004.shtml

Kitfield, James, "U.S. Response I: U.S. Ports and Waterways Remain Vulnerable," *Nuclear Threat Initiative*, July 2, 2002. As of August 3, 2005:
http://www.nti.org/d_newswire/issues/2002/7/2/1s.html#From

Klein, Gary A., *Sources of Power: How People Make Decisions,* Cambridge, Mass.: MIT Press, 1998.

L3 Communications Security and Detection Systems, "Vivid VIS 108," product specification, Woburn, Mass., October 2002. As of October 16, 2006:
http://www.bavakbeveiligingsgroep.nl/pdf/VIS_Eng_Span_Port_10_30_02.pdf

Lane, Earl, "Port Security: Concern Lingers Over New Scanners," *Newsday*, August 17, 2004, p. A18.

Li, Silas, and Tom O'Dwyer, "Computer Technology at Work: Ventilation Study for Times Square and Grand Central Stations," *Network*, Vol. 34, No. 10, Spring 1996. As of November 27, 2006:
http://www.pbworld.com/news_events/publications/network/issue_34/34_19_LiS_VentilationStudyTimesSq.asp

Libicki, Martin C., Peter Chalk, and Melanie Sisson, *Exploring Terrorist Targeting Preferences*, Santa Monica, Calif.: RAND Corporation, MG-483-DHS, 2007. As of June 28, 2007:
http://www.rand.org/pubs/monographs/MG483/

Liedtke, Michael, "Now Starring on the Internet: YouTube.com," *USA Today*, April 9, 2006. As of June 19, 2006:
http://www.usatoday.com/tech/news/techinnovations/2006-04-09-youtube-popularity_x.htm

Lloyd's Marine Intelligence Unit, *Lloyd's List, Ports of the World 2004,* London: Informa UK Ltd., 2003.

Long Island Area Scanning Resources, "Long Island Rail Road Frequencies," 2001a. As of August 3, 2005:
http://www.fordyce.org/scanning/frequencies/lirr.html

———, "M.T.A. Police Radio Codes," 2001b. As of August 3, 2005:
http://www.fordyce.org/scanning/scanning_info/lirrcode.html

Long Island Rail Road, "Douglaston," undated Web page. As of August 3, 2005:
http://lirr42.mta.info/stationinfo.asp?station=023

———, "Individual Station Timetables, Effective May 23–September 11, 2005," 2005a. As of August 3, 2005:
http://www.mta.info/lirr/html/ttn/lirrtt.htm

———, "Keeping Track," July 2005b. As of August 3, 2005:
http://www.mta.nyc.ny.us/lirr/KeepingTrack/

Long Island Rail Road History, "Port Washington Branch Stations," undated Web page. As of August 3, 2005:
http://www.lirrhistory.com/pwstas.html

Los Alamos National Laboratory, "lanl.arXiv.org e-Print archive mirror," undated Web page. As of August 3, 2005:
http://xxx.lanl.gov

Luczak, Marybeth, "Transit Security: What More Can Be Done?" *Railway Age*, September 2005. As of January 23, 2006:
http://findarticles.com/p/articles/mi_m1215/is_9_206/ai_n15674996

Luft, Gal, and Anne Korin, "Terrorism Goes to Sea," *Foreign Affairs*, November/December 2004.

Maersk Logistics, "Customs-Trade Partnership Against Terrorism (C-TPAT)," undated Web page. As of June 28, 2007:
http://www.maersklogistics.com/sw18198.asp

Magnusson, Paul, and Spencer Ante, "The Booming Security Business," *Business Week*, July 25, 2005, p. 33.

Massport, "Air Cargo Screening Tested at Logan; First Airport on Passenger Bag Screening Leads the Way on Air Cargo Scanning," press release, October 14, 2003. As of August 1, 2005:
http://www.massport.com/about/press03/press_news_airca.html

Mayer, Matt, "Updated Information on the Testing of Radiation Detection Equipment," ODP Information Bulletin No. 168, Washington, D.C.: Office of Domestic Preparedness, May 19, 2005. As of October 17, 2006:
http://www.ojp.usdoj.gov/odp/docs/info168.htm

Medby, Jamison Jo, and Russell W. Glenn, *Street Smart: Intelligence Preparation for the Battlefield for Urban Operations*, Santa Monica, Calif.: RAND Corporation, MR-1287-A, 2002. As of November 21, 2006:
http://www.rand.org/pubs/monograph_reports/MR1287/index.html

Metropolitan Transit Authority, "MTA Long Island Rail Road," 2004. As of November 24, 2006:
http://www.mta.nyc.ny.us/lirr/html/lirrmap.htm

MillenniaPartners, "Odyssey Logistics & Technology Receives Certification from the Customs-Trade Partnership Against Terrorism," press release, Danbury, Conn., August 24, 2004. As of October 17, 2006:
http://www.millenniapartners.com/News/NewsView.asp?NewsID=265

Mineta, Norman Y., "Statement of the Honorable Norman Y. Mineta, Secretary of Transportation, Before the Senate Subcommittees on Transportation, Committee on Appropriations and Surface Transportation and Merchant Marine, Committee on Commerce, Science, and Transportation, U.S. Senate," June 27, 2002. As of July 31, 2005:
http://testimony.ost.dot.gov/test/pasttest/02test/Mineta5.htm

Mitchell, Alan, "Panic on the PATH Train at 14th Street," *ScaryNY*, August 11, 2004. As of November 22, 2006:
http://scaryny.com/archives/2004/08/panic_on_the_pa.php

Moorman, Robert W., "Fire in the Belly?," *Homeland Security*, September 2004, p. 44.

Murakami, Haruki, *Underground*, New York: Vintage Press, 2001.

MU-Vision, Inc., *MU-Detector, a Novel Method of Detecting Nuclear Weapons, "Dirty" Bombs and Voids in Cargo*, undated white paper. As of August 3, 2005:
http://www.mu-vision.com/Composite%20MU-Detector%20Whitepaper_08-03-04.pdf

n2nov.net, "New York City Police Department Radio Signal Codes," October 19, 2005. As of November 22, 2006:
http://www.n2nov.net/nypdcodes.html

National Council for Science and the Environment, *CRS Reports: Transportation*, undated Web page. As of March 9, 2006:
http://www.ncseonline.org/NLE/CRS/Detail.cfm?Category=Transportation

National Fire Protection Association, and American National Standards Institute, *Standard for Fixed Guideway Transit and Passenger Rail Systems*, Quincy, Mass.: National Fire Protection Association, NFPA 130, 2003.

National Nuclear Security Administration, "Megaports Initiative: Protecting the World's Shipping Network from Dangerous Cargo and Nuclear Materials," undated Web page. As of March 14, 2006:
http://www.nnsa.doe.gov/megaports_initiative.htm

NationMaster.com, "Encyclopedia: Atlantic Avenue Tunnel," 2005. As of August 2, 2005:
http://www.nationmaster.com/encyclopedia/Atlantic-Avenue-Tunnel

Neubecker, Lt. Graig D., "Shark on Attack: America's First Armed Airborne Unit to Protect Our Coast," *Air Beat Magazine: Journal of the Airborne Law Enforcement Association*, September–October 2003. As of August 3, 2005:
http://www.alea.org/public/airbeat/back_issues/sep_oct_2003/shark_on_attack.htm

New York City Subway, "FAQ: Facts and Figures—Subway Technical FAQ," September 10, 2002. As of July 29, 2005:
http://www.nycsubway.org/faq/factsfigures.html

Northwest Airlines, "Find Station Information," undated(a) Web page. As of August 1, 2005:
http://www.nwa.com/services/shipping/cargo/products/station.shtml

———, "NWA Cargo Known Shipper Validation," undated(b) Web page. As of August 2, 2005:
http://www.nwa.com/services/shipping/cargo/about/shipper.shtml

"Nuke Detectors for LA-LB," *The Journal of Commerce*, June 7, 2005, p. 1.

Ogando, Joseph, "Hardened Air Cargo Container Relies on Careful Joining," *Design News*, November 18, 2002. As of August 2, 2005:
http://www.designnews.com/article/CA257378.html

Online Forum, "Could It Happen Here? Is the U.S. Prepared for a Chemical or Biological Weapons Attack?" February 11, 1998. As of July 29, 2005:
http://www.pbs.org/newshour/forum/february98/weapons_2-11.html

Orphan, Victor, Ernie Muenchau, Jerry Gormley, and Rex Richardson, *Advanced Cargo Container Scanning Technology Development*, paper presented at the 7th Marine Transportation System Research and Technology Coordination Conference, Washington, D.C., November 16–17, 2004. As of August 3, 2005:
http://trb.org/Conferences/MTS/3A%20Orphan%20Paper.pdf

Pardo, Shawn, "The Sonic-Anemometer May Save Lives, the Inventor Says," *The Word,* New York, Vol. 1, July 13, 1998.

Pauwels, Linda Pfeiffer, "From the Cockpit," October 2003. As of August 1, 2005:
http://www.alliedpilots.org/Public/Topics/Issues/air_cargo_safety.pdf

Panix, "This Is a Catch-All List of Assorted Frequencies for the New York Metro Area," undated Web page. As of July 29, 2005:
http://www.panix.com/clay/scanning/Frequencies/States/general.ny-metro.ny

Port of Los Angeles, "World Cruise Center," undated Web page. As of August 3, 2005:
http://www.portoflosangeles.org/recreation_Cruising.htm

———, "Security Measures Move Forward at the Port of Los Angeles," press release, San Pedro, Calif., November 10, 2004. As of October 17, 2006:
http://www.portoflosangeles.org/Press/REL_TSA%20Grant%20Acceptance.pdf

Port of Oakland, "Radiation Portal Monitoring (RPM) Program Implementation for U.S. Customs and Border Protection (CBP), US DHS," undated briefing. As of November 24, 2006:
http://aapa.files.cms-plus.com/SeminarPresentations/05_OpsIT_Borner-Brown_Jill.pdf

———, "Agenda: Regular Meeting of the Board of Port Commissioners," September 7, 2004. As of August 3, 2005:
http://portofoakland.com/pdf/boar_shee_040907.pdf

Public Law 108-90, An Act Making Appropriations for the Department of Homeland Security for the Fiscal Year Ending September 30, 2004, and for other purposes, October 1, 2003. As of August 1, 2005:
http://frwebgate.access.gpo.gov/cgi-bin/getdoc.cgi?dbname=108_cong_public_laws&docid=f:publ090.108

RadioReference.com, undated homepage. As of November 21, 2006:
http://www.radioreference.com

railfanpete, "M7 1 Pictures from New England Photos on Webshots," *Webshots*, December 29, 2004. As of November 25, 2006:
http://travel.webshots.com/photo/1239901528046352987GoyoYc

Rashbaum, William K., "Police Department Gives Officers a Guide for Detecting Terrorists," *The New York Times*, May 18, 2004, p. B7.

RFID Journal, "War Game Exposes Cargo Threat," January 13, 2003. As of July 29, 2005:
http://www.rfidjournal.com/article/articleview/265/1/66/

Roberti, Mark, "Target Tests RFID for Security," *RFID Journal*, December 10, 2004. As of October 17, 2006:
http://www.rfidjournal.com/article/articleview/1282/1/1/

Robin, Joshua, "Fighting Terror on the Subway, MTA Says Workers Are Trained to Report Suspicious Behavior," *New York Newsday*, July 13, 2005, p. A14.

Roboto, Philip, "Terrorism at Sea: U.S. Must Be Prepared," *PoliticsOL.com*, July 29, 2001. As of July 29, 2005:
http://www.politicsol.com/editorials/editorial-2001-07-29.html

Ross, Brian, "New Report Reveals Gaps in Port Safety," *ABC News*, October 13, 2004. As of October 17, 2006
http://abcnews.go.com/WNT/story?id=162480

Ross, Brian, Rhonda Schwartz, and David Scott, "Customs Fails to Detect Depleted Uranium," *ABC News*, September 11, 2002. As of November 24, 2006:
http://abcnews.go.com/WNT/story?id=129321&page=1

Rubacky, Tim, "How Safe Are We at Sea?" *CruiseMates*, undated. As of August 3, 2005:
http://www.cruisemates.com/articles/consumer/security.cfm

Safeboats International, LLC, "Collar," undated Web page. As of August 3, 2005:
http://www.safeboats.com/default/collar.html

Samuel Shapiro and Company, Inc., Consulting, "Customs-Trade Partnership Against Terrorism," undated Web page. As of August 3, 2005:
http://www.shapiro.com/./html/ctpat.html

Sanchez, Ray, "A So-So Tool Against Terror," *Newsday* (New York), March 10, 2005, p. A06.

Schneider, Greg, "Terror Risk Cited for Cargo Carried on Passenger Jets," *The Washington Post*, June 10, 2002, p. A01.

Schumer, Charles E., "Schumer: Madrid Attacks Should Be Wake-Up Call to Upgrade Rail and Subway Security," press release, March 14, 2004a. As of August 3, 2005:
http://schumer.senate.gov/SchumerWebsite/pressroom/press_releases/PR02501.madrid.031404.html

———, "Schumer Finds Fed's Proposed New Cargo Screening Rules Still Leave Gaping Hole in NY's Air Security," press release, November 28, 2004b. As of August 1, 2005:
http://schumer.senate.gov/SchumerWebsite/pressroom/record.cfm?id=264688&

———, "New Schumer Analysis of LIRR Stations Reveals Security Measures Are Painfully Inadequate," press release, March 7, 2005. As of August 3, 2005:
http://schumer.senate.gov/SchumerWebsite/pressroom/press_releases/2005/PR41511.LIRRand%20DHS.030705.html

Science Applications International Corporation, National Cooperative Highway Research Program, National Research Council, and American Association of State Highway and Transportation Officials, *Surface Transportation Security: A Field Personnel Manual*, Vol. 1: *Responding to Threats*, Washington, D.C.: Transportation Research Board, 2004.

"Security Boosted at LAX to Guard Against Missiles," *USA Today*, December 14, 2004. As of August 3, 2005:
http://www.usatoday.com/travel/news/2004-12-14-lax-missiles_x.htm

Security Management Online, "Library: Government," undated Web page. As of August 1, 2005:
http://www.securitymanagement.com/library/000132_gov.html

Simon, Herbert, "From Substantive to Procedural Rationality," in Spiro J. Latsis, ed., *Method and Appraisal in Economics*, Cambridge and New York: Cambridge University Press, 1976.

Smerd, Jeremy, "MTA Details Efforts to Detect Biochemical Attacks," *The New York Sun*, July 12, 2005. As of July 29, 2005:
http://www.nysun.com/article/16807

Smiths Heimann, homepage, January 1, 2005. As of August 1, 2005:
http://www.smiths-heimann.de/

Soares, Claire, "U.S. Developing Chemical Attack Sensor for Subways," Reuters, October 2001. As of July 29, 2005:
http://www.mercola.com/2001/oct/10/subway_chemical_sensor.htm

Spangenberg, Jr., Lt. Francis E., "The Job of a Transit Cop," undated. As of July 29, 2005:
http://www.nyc.gov/html/nypd/html/transportation/jobofa.html

Stana, Richard M., *Container Security: A Flexible Staffing Model and Minimum Equipment Requirements Would Improve Overseas Targeting and Inspection Efforts*, Washington, D.C.: U.S. Government Accountability Office, 2005a.

————, *Homeland Security: Key Cargo Security Programs Can Be Improved: Testimony Before the Permanent Subcommittee on Investigations, Committee on Homeland Security and Governmental Affairs, United States Senate*, Washington, D.C.: U.S. Government Accountability Office, May 26, 2005b. As of November 22, 2006:
http://purl.access.gpo.gov/GPO/LPS61823

Staten, Clark, "Chemical Attack—Are We Prepared?" Emergencynet News Service, May 22, 1995. As of July 29, 2005:
http://www.emergency.com/chemattk.htm

Stevens, Donald, Terry Schell, Tom Hamilton, Richard Mesic, Michael Scott Brown, Edward W. Chan, Mel Eisman, Eric V. Larson, Marvin Schaffer, Bruce Newsome, John Gibson, and Elwyn Harris, *Near-Term Options for Improving Security at Los Angeles International Airport*, Santa Monica, Calif.: RAND Corporation, DB-468/1-LAWA, 2004. As of August 3, 2005:
http://www.rand.org/pubs/documented_briefings/DB468-1/

Stoller, Gary, "Airliners May Get Missile Defenses," *USA Today*, July 14, 2005, p. 1A. As of August 2, 2005:
http://www.usatoday.com/news/nation/2005-07-13-airliners-defenses_x.htm

Stone, David M., "Testimony of David M. Stone, Assistant Secretary of Homeland Security, Transportation Security Administration, Department of Homeland Security, on 9/11 Commission Recommendations of Civil Aviation Security Before the Subcommittee on Aviation Committee on Transportation and Infrastructure, United States House of Representatives," August 25, 2004.

Strengthening the Global Partnership, "AAAS Expert Proposes 'Layered Defense' to Protect Against Smuggled Nuclear Materials," June 23, 2005. As of August 3, 2005:
http://www.sgpproject.org/Personal%20Use%20Only/AAASLayeredDefense.html

Stryker, Robin, "Beyond History Versus Theory: Strategic Narrative and Sociological Explanation," *Sociological Methods & Research*, Vol. 24, No. 3, 1996, pp. 304–352.

Telair International, "Federal Aviation Administration Approves Telair International's Blast Resistant Aircraft Baggage Container," press release, Rancho Dominguez, Calif., February 8, 2002. As of November 24, 2006:
http://www.telair.com/02-01News/02Feb8-A.html

Thomas, Timothy L., "Information-Age 'De-Terror-ence,'" *Military Review*, January–February 2002. As of October 15, 2006:
http://usacac.army.mil/CAC/milreview/English/JanFeb02/JanFeb02/thomas.pdf

————, "Al Qaeda and the Internet: The Danger of 'Cyberplanning,'" *Parameters*, Spring 2003. As of July 28, 2005:
http://carlisle-www.army.mil/usawc/Parameters/03spring/thomas.htm

Tirschwell, Peter, "A Chat with the Director of C-TPAT," *The Journal of Commerce*, January 2, 2006, p. 1.

Trager, Rachel, "Subway Riders Scrutinize Security," *Columbia Daily Spectator*, April 27, 2005.

Transportation Research Board of the National Academies, "NCHRP 20-59 [Active]: Surface Transportation Security Research," undated(a) Web page. As of November 21, 2006:
http://www.trb.org/trbnet/projectdisplay.asp?projectid=644

————, "TCRP J-10 [Active]: Public Transportation Security Research," undated(b) Web page. As of November 21, 2006:
http://www.trb.org/trbnet/projectdisplay.asp?projectid=1186

Transportation Security Administration, U.S. Department of Homeland Security, "Traveling with Pets," undated Web page. As of November 24, 2006:
http://www.tsa.gov/travelers/airtravel/assistant/editorial_1036.shtm

————, "Air Cargo Strategic Plan," Washington, D.C., November 17, 2003.

————, "TSA Takes Another Step to Protect At-Risk Air Cargo," press release, April 9, 2004a. As of August 2, 2005:
http://www.tsa.gov/press/releases/2004/press_release_0409.shtm

————, "TSA-ASAC Cargo Working Groups Final Recommendations of the Working Group," December 14, 2004b. As of November 22, 2006:
http://www.tsa.gov/assets/pdf/ASAC_FAS_WG_Recommendations_121404.pdf

United Nations Office on Drugs and Crime, "Conventional Terrorist Weapons," undated. As of August 3, 2005:
http://www.unodc.org/unodc/terrorism_weapons_conventional.html

US Airways, "About Cargo," undated Web page. As of November 24, 2006:
http://www.usairways.com/awa/content/traveltools/cargo/default.aspx

U.S. Coast Guard Sector Los Angeles–Long Beach, undated homepage. As of November 25, 2006:
http://www.uscg.mil/d11/sectorlalb/index.htm

U.S. Code, Title 33, Chapter I, Subchapter P, Part 165.1154, Security Zones; Cruise Ships, San Pedro Bay, California, July 1, 2003.

U.S. Customs and Border Protection, "C-TPAT Frequently Asked Questions," undated Web page. As of October 17, 2006:
http://www.cbp.gov/xp/cgov/import/commercial_enforcement/ctpat/ctpat_faq.xml

————, "National Targeting Center Keeps Terrorism at Bay," *U.S. Customs and Border Protection Today*, March 2005. As of October 17, 2006:
http://www.cbp.gov/xp/CustomsToday/2005/March/ntc.xml

U.S. Department of Defense, *Exhibit R-2, RDT&E Budget Item Justification*, February 2003. As of August 3, 2005:
http://www.dtic.mil/descriptivesum/Y2004/OSD/0604618D8Z.pdf

———, *Report of the Defense Science Board Task Force on Preventing and Defending Against Clandestine Nuclear Attack*, Washington, D.C.: Office of the Under Secretary of Defense for Acquisition, Technology, and Logistics, June 2004. As of November 24, 2006:
http://purl.access.gpo.gov/GPO/LPS58237

U.S. Department of Homeland Security, "Department of Homeland Security Announces New Measures to Expand Security for Rail Passengers," press release, May 20, 2004. As of November 24, 2006:
http://www.dhs.gov/xnews/releases/press_release_0412.shtm

———, *Performance Budget Overview*, Appendix A: *Verification and Validation of Measured Values*, Washington, D.C.: DHS, FY2006.

U.S. Department of Transportation, *Report to Congress: Aviation Security Aircraft Hardening Program*, Washington, D.C., December 1998.

U.S. General Accounting Office, *Aviation Security: Vulnerabilities and Potential Improvements for the Air Cargo System*, Washington, D.C.: GAO, GAO-03-344, December 2002. As of November 24, 2006:
http://purl.access.gpo.gov/GPO/LPS32172

———, *Nonproliferation: Further Improvements Needed in U.S. Efforts to Counter Threats from Man-Portable Air Defense Systems*, GAO-04-519, May 2004. As of November 21, 2006:
http://www.gao.gov/new.items/d04519.pdf

U.S. Government Accountability Office, *Preventing Nuclear Smuggling: DOE Has Made Limited Progress in Installing Radiation Detection Equipment at Highest Priority Foreign Seaports*, GAO-05-375, March 2005a. As of March 14, 2006:
http://www.gao.gov/new.items/d05375.pdf

———, *Container Security: A Flexible Staffing Model and Minimum Equipment Requirements Would Improve Overseas Targeting and Inspection Efforts*, GAO-05–557, April 2005b. As of August 3, 2005:
http://www.gao.gov/new.items/d05557.pdf

———, reports and testimony, March 8, 2006.

U.S. House of Representatives, *National Intelligence Reform Act of 2004*, H.R. 5223, 108th Congress, 2d Session, October 5, 2004a. As of November 22, 2006:
http://thomas.loc.gov/cgi-bin/query/z?c108:H.R.5223.IH:

———, *House Report 108-541, Department of Homeland Security Appropriations Bill, 2005*, June 15, 2004b. As of November 21, 2006:
http://frwebgate.access.gpo.gov/cgi-bin/getdoc.cgi?dbname=108_cong_reports&docid=f:hr541.108.pdf

———, *Department of Homeland Security Appropriations Bill, 2006*, May 13, 2005. As of August 1, 2005:
http://thomas.loc.gov/cgi-bin/cpquery/T?&report=hr079&dbname=cp109&

U.S. Library of Congress, *Department of Homeland Security Appropriations Bill, 2005*, House Report 108-541, June 15, 2004. As of July 31, 2005:
http://thomas.loc.gov/cgi-bin/cpquery/?&db_id=cp108&r_n=hr541.108&sel=TOC_428137

Vause, John, "Missile Defense for El Al Fleet," *CNN.com*, May 24, 2004. As of August 2, 2005:
http://www.cnn.com/2004/WORLD/meast/05/24/air.defense/

WABC, "More Police in NYC Subway," July 10, 2005. As of November 22, 2006:
http://abclocal.go.com/wabc/story?section=news&id=3237248&ft=print

Wallace, Kelly, Rob Marciano, Suzanne Malveaux, Mary Snow, Chad Myers, and Sean Callebs, "Rove Leaked; No Tunnel Cells; Cleaning Up; Return to Space," *CNN Daybreak*, July 12, 2005.

Walsh, Don, "Tourism and Terrorism: A Difficult Journey Ahead for the Cruise Ship Industry," *Navy League of the United States*, December 2002. As of August 3, 2005:
http://www.navyleague.org/sea_power/dec_02_51.php

Wein, Lawrence, Alex Wilkins, Manas Baveja, and Stephen Flynn, "Preventing the Importation of Illicit Nuclear Materials in Shipping Containers," *Risk Analysis*, Vol. 26, No. 5, October 2006, pp. 1377–1393.

Weiss, Murray, and Clemente Lisi, "The Tracks of Our Fears: NYPD in Subway 'Bomb' Sweeps," *The New York Post*, March 22, 2004, p. 5.

William Nicholas Bodouva and Associates, "Rail Projects," undated Web page. As of November 27, 2006:
http://www.bodouva.com/rail/index.html

Willis, Henry H., and David S. Ortiz, *Assessing Container Security: A Framework for Measuring Performance of the Global Supply Chain*, Santa Monica, Calif.: RAND Corporation, RB-9095-RC, 2005. As of November 21, 2006:
http://www.rand.org/pubs/research_briefs/RB9095/index1.html

Wilson, Jan, "TSA Focuses Increased Attention on Air Cargo," *GSN: Government Security News*, undated Web page. As of August 1, 2005:
http://www.gsnmagazine.com/aug_04/cargo_screening.html

WirelessAdvisor.com Forums, "Northeastern US Wireless Forum: Reception in NYC Subway," undated(a) Web forum post. As of August 3, 2005:
http://forums.wirelessadvisor.com/showthread.php?t=4341

―――, "Wireless Coverage on Subways," undated(b) Web forum. As of August 3, 2005:
http://forums.wirelessadvisor.com/archive/index.php/t-863.html

WorldSecurity-index.com, "Science Applications International Corporation," undated(a) Web page. As of October 16, 2006:
http://www.worldsecurity-index.com/details.php?id=337&lang=EN

―――, "Thermo Electron Corporation," undated(b) Web page. As of October 16, 2006:
http://worldsecurity-index.com/details.php?id=1031

―――, "Products/Services—Non-Intrusive Large Scale—KNM Media LLP," 2001. As of August 3, 2005:
http://worldsecurity-index.com/cats.php?cat=1350

Young, John, *Cryptome*, undated Web site. As of November 21, 2006:
http://www.cryptome.org